INSPIRED BY LIGHT

© RIBA Publishing, 2022

Published by
RIBA Publishing
66 Portland Place
London W1B 1AD

ISBN 978-1-85946-905-7

Reprinted 2023

British Library Cataloguing-in-Publication Data
A catalogue record for this book is available
from the British Library.

Commissioning Editor: Elizabeth Webster
Assistant Editor: Clare Holloway
Production: Jane Rogers
Designed by CHK Design
Printed and bound by
Short Run Press Limited, Exeter
Cover image: Interior Designer: Designers Guild
Photographer: Breed Media.
Architect: Stephen Marshall Architects

While every effort has been made to provide
technical guidance, this information is only a
guide. Regulations change and are different
from country to country and must in all cases
be checked locally. It is strongly advised that
a qualified electrician is consulted for the
installation of all light fixtures. Neither the author
nor the publisher shall be liable for any claims.

www.ribapublishing.com

INSPIRED BY LIGHT

A DESIGN GUIDE TO TRANSFORMING THE HOME

SALLY STOREY

ACKNOWLEDGEMENTS

I am deeply grateful to everyone who has allowed me to use photographs of their houses without which this book would not have been possible. I would like to thank the architects and designers with whom I worked to achieve the wonderful results; they are the heroes and my work is to enhance their creations with light.

On a personal note, my thanks goes to Cara Salmon for her outstanding work, energy and enthusiasm in collating the image library and supporting me throughout. This book would not have been completed without the dedication and help of my personal assistant, Alice Berry, Sophie Pither's brilliant editorial support and the contribution of my husband Christopher Fordham.

ABOUT THE AUTHOR

Sally Storey, Founding Director of Lighting Design International and Creative Director of John Cullen Lighting, is one of the UK's leading lighting experts. Her extensive knowledge and experience have led to her travelling all over the world designing lighting schemes. She is the recipient of numerous industry awards and recently received worldwide recognition with the lifetime achievement award from LIT Awards based in Los Angeles.

TABLE OF CONTENTS

PART 1 TECHNICAL ASPECTS OF LIGHTING

PART 2 LIGHTING THE HOME

PART 3 CASE STUDIES

	Introduction	VIII
01	How lighting can support the architectural scheme	2
02	Principles and practices of lighting design	10
03	Lighting techniques	16
04	Working with LED	30
05	Controlling your scheme	40
06	How to create a lighting plan	48
07	Front doors and first impressions	58
08	Staircases	70
09	Living rooms	82
10	Dining rooms	92
11	Kitchens	100
12	Bedrooms and dressing rooms	110
13	Bathrooms	120
14	Unusual spaces	130
15	Leisure rooms and pools	140
16	Contemporary terraced house	152
17	Traditional terraced house	160
18	Barn conversion	168
19	Alpine chalet	166
20	Contemporary villa	186
21	Duplex penthouse	196
	Glossary of terms	204
	Index	206
	Image credits	208

INTRODUCTION

The creative and technical art of lighting design may be relatively new, but appreciating light is as old as human existence. The play of sunlight on everything around us is a daily inspiration. Dappled sunlight through rippling leaves, rays glinting off raindrops, long shadows on a late afternoon, warming bright beams across polished floorboards – natural light affects our mood and visual appreciation of our surroundings. Modern lighting design re-creates drama in our living spaces, highlighting and illuminating and using a combination of light and shadow to enhance the architecture and appeal of our homes.

Theatre directors have long understood the potential for light to create atmosphere and engage an audience. Powerful combinations of light and shadow, tone and colour help create the mood for each set, directing our attention and playing with our expectations and emotions. This same appreciation and understanding can be applied to lighting architecture – traditional or contemporary – creating a dialogue between space and light, allowing an interior to be revealed or concealed as the day progresses. In our homes, we are the audience to our lighting design. Lighting designers make the theatre of light work both practically and aesthetically.

For centuries, ambience in interiors was created by the use of layered light – oil lamps or candles, torchères on walls and firelight in the hearth. From the 19th century, gaslight created pools of atmospheric light in our streets and homes. The arrival of the tungsten filament bulb in the 1920s removed the subtlety of these layered sources. A tungsten pendant in the centre of the room often became the single light source, providing easy, inexpensive and clean light. Flooding rooms with this even light came at a cost to ambience. The adoption of early energy-efficient tubular fluorescent lamps, with their flat, cold light, killed any remaining layered atmosphere in the home.

Figure 0.1

Shadows are as essential an element of lighting design as natural light. The screening on the window filters natural light; as the day progresses, so the shadows change.

The drive for energy efficiency was commendable, but the lighting was harsh, would flicker and was difficult to dim. The tide turned with the arrival of the tungsten halogen lamp in the late 1970s. At last, lighting designers had a miniature accent tool: the size of a halogen capsule was smaller than traditional tungsten sources. This allowed a smaller reflector to control the light, and a more subtle approach everywhere, from homes to hotels, restaurants and museums. Downlights and spotlights proliferated.

A leap forward for energy efficiency came in the 1990s with the white LED (light-emitting diode) light source. It gradually replaced tungsten halogen reflector lamps and ended the compact tungsten filament source of the traditional bulb. Unfortunately, these first LED light sources offered only a cool colour temperature with poor colour rendition. LED technology has now caught up, to provide a better spectrum of colour and light quality, and lenses to direct light. Today, these LEDs are the light source of choice, offering wonderful scope for lighting designers.

Think of a favourite south-facing room on a late summer's afternoon: light pooling on the floor, shadows coming and going, diffuse sunlight filtering through windows. The south-facing room has energy, whereas east-facing rooms, after morning and without artificial light, are shrouded in gloom. You will avoid them if you can, waiting until the following morning, when a new day brings warmth and light into these rooms. Few of us now rise with the sun and rest from dusk. However, we still reference the position of the sun throughout the day, knowing at an elemental level where we are in the diurnal round, and how its light and shade affect our surroundings and shape our emotions. Successful lighting creates homes which people can enjoy after daylight fades.

Our workplaces are lit at a cool light temperature, with a bright, uniform ambient level and little contrast. Conversely, we want our homes to have a more relaxed style. Lighting designers visualise how houses transition from dawn to dusk, supplementing and accenting through the day, until artificial light takes over entirely.

An architect, when designing a house, will visualise a project in three dimensions from the outset. The design will flow from the shape and structure of the space, the choice of materials and the role of natural light. The architect will consider natural light from windows, doors, skylights, apertures, glass walls and ceilings, which will in turn shape the visual appearance of the house, internally and externally.

Intelligent and creative lighting design reinforces the architect's intent. Light can be applied to emphasise volume, highlight interesting structures and assist with flow through a building. It can be manipulated to create balance between disparate spaces, to differentiate between materials and to create drama. The play and control of light colour and the amount of emphasis and focus created influence mood within a space. It is a collaborative effort to deliver the vision of the architect and interior designer. This book illustrates how inspiring lighting design is achieved, and how light can enhance the architect's and interior designer's vision to make a building work better, be more pleasurable to inhabit and more beautiful.

An architect plays with volumes and materials, an interior designer with finishes and texture, a lighting designer with layers of light to enhance the design of the architect and interior designer. We have many options in our artificial lighting toolbox for building up layers of light, to create beautifully lit spaces and intangible benefits of wellbeing and emotions of calm or excitement. We provide task lighting for reading, cooking or working at a desk. We consider flow and overall feel, allowing all areas of a house to be enjoyed in different ways at different times of

Figure 0.2a
Figure 0.2b
Figure 0.2c

*The same bathroom, pictured
in the morning (a), afternoon, with
light from the west filtering through
shutters (b), and evening (c).*

day or night. Lighting should add an extra dimension to make the very best of the space, creating depth and height, inviting corners, bright areas and sometimes shadows and contrasts, and focusing attention on important features. It's about the balance of light and shade and bringing new energy to an interior. Like architects, lighting designers consider the entire house – inside and out.

Part 1 of this book discusses the three main elements of lighting: ambient, task and accent, and how these must be considered and balanced in each room and connecting area. Lighting solutions may be repeated in rooms, but the interpretation of the lit space will alter; understanding the effects is the skill. Knowing how many millimetres from the wall a linear LED should sit, and how to conceal it, how to diffuse it and how to balance it with other lighting is all part of the job of the lighting designer. Part 2 illustrates solutions for all types of homes to give architects, interior designers and homeowners ideas and inspiration for their projects. I have illustrated key lighting techniques which work well in different parts of a home. The main rooms are covered but entire books could be written on specialist subjects like lighting art and joinery, which are just touched upon here. In Part 3, I have chosen a selection of beautiful houses with different architectural styles to illustrate a range of lighting design techniques and approaches. Each case study walks the reader through the home, illustrating how a lighting theme works in a single residence, how it complements the architecture and enhances the building and décor.

The joy of my job as a lighting designer is to be continually inspired. Each new space presents its own wonder, challenges and excitement. I hope that some of these ideas in different spaces will also uplift you. My aim is to inspire, intrigue and illuminate.

PART 1

TECHNI
ASPECT
OF LIGH

CAL S TING

CHAPTER 1

HOW LIGHTING CAN SUPPORT THE ARCHITECTURAL SCHEME

Good lighting does not steal the show – it makes heroes of architecture and interiors. It changes the environment we inhabit and determines how we perceive space. Playing with light changes the character, look and mood of a room. Doing it well is an art.

EARLY INTEGRATION OF
THE LIGHTING SCHEME

Figure 1.1

The stone wall in the rear is lit by multiple sources, with narrow beams top and bottom. This is achieved with individual sources but can also be arranged in a linear profile with special optics. In contrast, a diffused linear LED is concealed under the counter and gives a very soft, even light onto the stools.

Commissioning a lighting designer at concept stage, when the architect or interior designer has developed first plans with furniture layouts, will mean a harmonious and less costly build. There is a lot of electrical know-how to be considered, which affects construction. Lighting designers should have expertise in both.

Our first lighting questions include: How will spaces work at different times of day and night? How much light will be needed? And how will the user inhabit the space?

The lighting needs to be visualised both in plan form and three dimensions. When we enter a room, our initial impressions come from where our eyes are drawn. The lighting designer can control this, choosing what to light prominently. Where the light falls is where our eye will follow. This is why we opt for a selection of lighting techniques, each with its own purpose. Decorative lighting might first catch our eye – a chandelier or pendant perhaps, a lit picture, an illuminated fire surround or a well-lit reading chair. Placing architectural light fittings with a view to where the light will fall, and be reflected, creates the greatest effect.

A lighting layout to avoid is the standard grid of downlights. This rarely bears any relationship to how furniture is arranged or the elevations, so ends up merely washing flat light onto the floor. Your eye is drawn to the floor and outstanding architectural features dissolve into the background. Decide first where light should be directed; then make the lighting plan.

Best results are achieved when a lighting designer appreciates the architectural vision from the outset so that the lighting can be fully integrated into the fabric of the building. With contemporary buildings, the lighting is often concealed within a recessed detail in ceilings, walls and joinery. If considered at the outset, the lighting designer can draw the details for the architect to ensure correct dimensions are used to deliver the best lighting solution, which is always better than installing lights as an afterthought.

Figure 1.2

By day, the glorious glass blocks sparkle in natural light from behind.

Figure 1.3

By night, an LED with narrow-grazing optics is recessed in front of the glass blocks to shaft light down the inside. This highlights the white grouting and the glass blocks appear grey. The desk is lit from above by two downlights and below by an LED strip.

Wiring needs to be installed early on, at the same stage as the plumbing, so planning is essential. The lighting designer can help encourage decisions on furniture layouts to ensure lights are positioned to allow a variety of effects. Electrical supplies must be planned early, so that the lighting can be developed as the detail of the design progresses. Allowing for early flexibility of supplies can save costly rewiring later.

MANIPULATING
VOLUME

Figure 1.4

This long corridor is foreshortened by the play of light uplighting each bay and introducing simple downlight on the rug. Focus is created with the picture lit at the end and a lamp.

Figure 1.5

The corridor has no ceiling or floor lights; instead, uplights and downlights are built into slots recessed in the wall, creating a contemporary form of wall light, leading to the kitchen beyond.

Figure 1.6

In this classical house, uplights emphasise the arches. Wall lights provide the general light and, with downlights, create soft pools of light on the floor.

Lighting a house can be transformative, influencing how spaces are read, used and enjoyed. It can influence how the volume of a space is appreciated. For instance, by lighting ceilings and walls, the sense of space is increased. Conversely, the room will seem smaller and more intimate if only its centre is lit.

Deciding how to draw one's eye to a feature is achieved by controlling light. An object may be top-lit, cross-lit or even backlit, and each effect will describe the object differently. The best choice will depend on the object, the room and how its surroundings are lit.

HIGHLIGHTING ARCHITECTURAL FEATURES AND MATERIALS

Figure 1.7 a
Figure 1.7 b
Figure 1.7 c

Ambient light on the texture gives a flat appearance (a). The linear light is offset by 150mm, which starts to show the texture (b). Narrow-beam uplights close to the surface emphasise the texture (c).

Specific features, such as arches and windows, can be highlighted and enhanced by the application of light.

Using light close up (we call this close offset) is a useful device when a wall has texture. However, it can be a problem with a flat plastered wall that seems smooth when lit from a distance, but when lit up close can appear to be covered in defects.

RELATING INTERIOR AND EXTERIOR SPACES

Lighting has a vital role in the relationship between indoor and outdoor spaces. In daytime, natural light will draw the eye from the interior to the exterior, providing an extension of the space. Lighting can do the same at night, effectively doubling the field of vision. Where a site is constrained, light can visually enlarge the space. The outside needs to be brighter than the inside and the area in front of the window needs to be underlit to draw the eye out. In another instance, there may be nothing to light, and to avoid the blackness of glass at night, light could be integrated into the ceiling or pelmet to wash down the curtains. This gives a definitive end to the room and closes off the space, creating a more intimate space.

Figure 1.8

The lighting inside is deliberately less intense beside the glazing to avoid reflection and ensure the outside is brighter at night. The downlight in the canopy lighting the terrace draws the eye outside. The trellis lighting is mounted on 30cm brackets that let the light skim down the planting, so there is interest from both downlight and uplight. This technique works well for green walls; the length of bracket depends on the depth of growth on the wall. Here, it also provides more light to the outdoor dining area as light reflects off the limestone paving.

DEFINING SPACES

Figure 1.9

This open-plan room is made up of a bar, kitchen and dining area. Each lighting solution is individual to each area, and in between them a play of shadow creates virtual divisions.

Lighting can be used to re-create rooms within a large open-plan area. In an open-plan kitchen/dining/living room, the activity in the kitchen area is different to the relaxed seating/TV area or the dining area; the dining area may be used for evening dining or as homework space or reading spot during the day. Each purpose needs delineation in the lighting design plan.

CHAPTER 2

PRINCIPLES AND PRACTICES OF LIGHTING DESIGN

When starting a project, the brief is key. Working effectively with both the client and design team is critical to achieve the desired outcome. Early in the design process, the lighting designer should highlight the spaces where focus is required, where to give accent or to leave in shade, what tasks are to be performed, and then ensure that the desired lighting strategy is possible.

Figure 2.1

In this dark room (a), light directs the eye to points of interest. These include the uplit fireplace and the flowers that come to life under the narrow-beam (10°) spotlight above the coffee table (b). This contrasts with the soft ambience created by the uplight on top of the bookcases. Downlights are used only as accent lighting directed towards the bookcase and the narrow beam onto the coffee table.

When we look at an object or surface, we are interpreting the light that is reflected from it. Our eyes respond to light, changing, adapting and manipulating it, reducing or increasing contrast. Lighting design harnesses light to manipulate and transform space.

Figure 2.2a
Figure 2.2b
Figure 2.2c

Lighting design transforms a space, as these three examples of the same corridor demonstrate. When the corridor is lit only with downlights (a), the walls remain dark and the narrow pools of light create focus and a transition through the space. The corridor appears different when the walls are uplit (b), providing reflected light; there is no focus and the space seems larger and taller. The use of both uplights and downlights (c) brings the best effect. The mood can be manipulated by adjusting the effects of downlights and uplights separately.

LAYERING OF
LIGHTING EFFECTS

Figure 2.3

In this interior, layers of light are created by perimeter ambient light (making the ceiling appear to float from the wall) and the soft light under the bench, combined with carefully positioned downlights.

Architects and interior designers employ a mixture of textures and finishes in many ways, combining shapes, surfaces, materials, fabrics, colours and textures to build up the palette of a project. A successful lighting scheme will have many effects which work together to enrich the whole: downlights, uplights, linear lights, spotlights, step lights, decorative lighting (table lamps, pendants, wall lights), floor washers, picture lights. Each has a function, but it is the combination which creates the magic. The skill is layering the lighting effects, allowing space to be transformed by changing the emphasis at will and considering how objects change under different lighting. The luminance of every light is important. This is the measure of luminous intensity – how much light passes through, or is emitted by or reflected from, a particular area or surface. Light is about more than being able to see – it's about brightness, colour and intensity, and responding to a space and to the materials to be lit.

THREE CONSTITUENT PARTS OF A SCHEME: AMBIENT LIGHTING, ACCENT LIGHTING, TASK LIGHTING

In developing a lighting scheme, there are generally three types of light to consider: ambient, accent and task. They work in harmony: varying their intensity changes the functionality and atmosphere of a room. Other considerations will govern choices, such as the constraints of the site, the architectural expression and a traditional or contemporary design, but the three basic constituent parts of a lighting plan remain the same.

Ambient lighting is the general background lighting, akin to natural daylight on an overcast day. It does not bring focus to the room but allows the room to be seen and used.
Accent lighting brings focus and directs the eye to a particular place, just as a ray of sunlight shining on an object on an otherwise overcast day will bring life and energy to that object.
Task lighting shines usually bright light onto areas used for daily tasks – preparing and cooking food, working at home or shining onto a favourite reading spot.

I strive to create effects by using shadow and contrast. We are in awe of nature when the light is dynamic and changing. Avoid the deadening effect of a flat, uniform light. Use ambient lighting as the base of your canvas and play with focus and accents to create interest and dynamism.

An overcast sky produces a flat, even, shadowless environment, in contrast to the effect of an overhead sun on a clear day. Strong light can bleach out colour, but it creates dramatic deep shadows which move as the sun appears to roll across the sky. The effects of light in nature influence lighting design, as changing light brings different responses. Dynamic light effects such as sunshine reflecting off moving water are mesmerising, as is the effect of reflection – the water becoming a mirror. These natural light effects during daylight can be just as strong at night with imaginative lighting.

Figure 2.4a
Figure 2.4b

*In the staircase of this converted convent, natural light
changes throughout the day, from morning (a) to evening (b).*

CHAPTER 3

LIGHTING TECHNIQUES

We need to bear in mind how buildings and interiors should be lit for best effect both during the day and after dark. This will start with the function of a room, and the spaces within it. Most spaces require some background – ambient – light as well as lighting for specific task areas, and areas of interest that will benefit from accent lighting. This chapter explores the techniques available to create this lighting.

Figure 3.1

Narrow-beam spotlights at high level light the art pieces. The effect is accentuated by the dark background. A wide beam lights the wall behind the chairs.

INDIRECT AMBIENT LIGHTING

Uplighting

This technique uses the ceiling as a reflector, which diffuses the light. Cove lighting, described below, is a form of uplight, as is placing LED strips on top of cabinets to reflect light off the ceiling. The effect is simple diffuse light and can reduce the number of direct downlight sources.

Figure 3.2

Uplighting above the kitchen cabinets and in the ceiling cove provides a bright, ambient light. This creates a white light during the day and a warmer effect in the evening. Pendants over the island and low-glare downlights between the pendants create task lighting on the island. A soft LED underlight below the counter catches the bar stools and provides an additional layer of accent light. While the kitchen is bright, this brightness is created by a build-up of layers of light.

Figure 3.3

A soft linear light in the cove illuminates the perimeter of the space; the depth is 150mm from the ceiling so the centre is still dark. There is another uplight from the dropped ceiling of the room beyond. The output of this is greater and the intensity higher.

Cove and Coffer Lighting

Cove lighting is a technique that directs light to a ceiling plane from a cove to provide general diffuse illumination. It is useful for providing ambient light at high level. It is usually achieved by an LED strip with a specific colour temperature and intensity. I recommend placing the LED strip in an extrusion with an opal diffuser to ensure individual dots of light are uniform. Or you could use a continuous homogenous flexible LED strip, which works well with curves. There are several varieties of extrusion – I often use a quadrant or, if a square extrusion is selected, I set this at an angle to maximise the spread of light in a cove. The ceiling's linear light needs to be level with the upstand of the cove to avoid harsh shadows. The evenness of the light will depend on the distance to the ceiling: 100mm is usually the minimum. A coffer in architecture is similar to a cove except that it is always sunk into a ceiling.

Cove and coffer details and finishes are important to achieve the desired effect. A small, shallow cove close to the ceiling will provide a strong line of light, whereas a deep cove further from the ceiling gives a softer, more even effect. The light from a cove is usually soft and provides a shadow-free general light if installed correctly. It can also combine colour temperatures to have a cooler daytime light and a warmer evening light. One should try to have a matt finish to avoid reflection of the light source.

Figure 3.4a
Figure 3.4b

This corridor is dull with just the uplights in the openings (a). The general light of the coffer transforms the space (b), allowing the uplights to provide an interesting architectural play on the openings.

Figure 3.5

A linear LED is concealed in a recess in the ceiling at the top of the wall. The light is more intense at the top as there is a small offset from the wall. The greater the distance from the wall, the softer the linear line of light will be.

Perimeter Wash, Grazing and Wall Washing

These types of lighting use perimeter walls to provide indirect, ambient light.

Perimeter Wash

Perimeter wash lighting can be achieved from an LED strip in a shadow-gap cove around the edge of a room, directing light downwards. Colour and intensity can vary with the specification of the LED and the shadow-gap detail. For example, if the gap is too small, the line of light will be too strong. If there is at least 40–50mm from the wall, the effect is softer.

Where the LED is positioned is important. Be careful with the viewing angles to ensure the light source is invisible. If the cove is concealed at both ends and one cannot easily look into it. The LED strip can be positioned to light directly down instead of reflecting off the ceiling; this direct solution is best to maximise the effect down the wall. Always ensure you conceal the source of the LED light strip. The indirect solution shines onto the top of the recess and reflects back and there is no direct view of the light source, but less light. This is necessary when you have a direct view into the lighting slot. Also, consider the finish of the wall: a gloss finish can reflect the light source, so use the indirect solution; a matt finish is more sympathetic and works better with direct light.

Grazing

Grazing is used to add interest to a textured wall and provide ambient light. This is achieved either with multiple individual LEDs or a linear LED strip with a narrow-beam (10°) optic and spreader lenses (which elongate the beam) usually located in a perimeter slot. To avoid seeing the light source, a custom detail needs to be created. This can mean the ceiling shields half the light source and half is direct. This can be used when there are direct viewing angles as the ceiling projection helps conceal the LED. The interior of the recess should be painted white to reflect light and increase infill light at the top of the wall. The effect sends more light down the wall than using a diffuse LED strip. It is not ideal if pictures are hung on the wall, as there would be large shadows from anything projecting from the wall. The same technique can be used as an uplight to a wall.

Figure 3.6

In this central courtyard, a recessed linear grazer uplights the stone wall and silhouettes the planting. This contrasts with the individual uplights between the timber fins.

Figure 3.7

Track fixtures (out of view) wall wash the cabinets, providing reflected general light.

Figure 3.8

If using a scallop light effect (an arc of light on a wall), the light shapes must line up with vertical detailing on the walls, such as these cupboard doors.

Wall Washing

Two types of wall washing are described below.

The first is recessed directional ceiling lights, offset from the wall, with a wide-beam light source and sand-blasted lenses, which provide a soft, even wash on the wall. Here they create a scallop of light. The effect is an indirect soft reflection off the wall, so any texture will be gently flattened rather than exaggerated. The scallops of light must line up with features on the wall, such as cupboard doors or windows, or the appearance will seem unconsidered.

The second wall wash type can be created from recessed or track-mounted fixtures with special reflectors or optics, designed to give an even, flat light. This solution is often adopted by art galleries to light pictures with an even wall wash rather than highlight individual pieces. The offset and separation of these fixtures are determined by the ceiling height. Adequate glare control is important when viewing side on, as many products work well head-on but can be glaring when viewed from the side. In this respect, the fixture should have some louvres to prevent side glare.

DIRECT AMBIENT LIGHTING

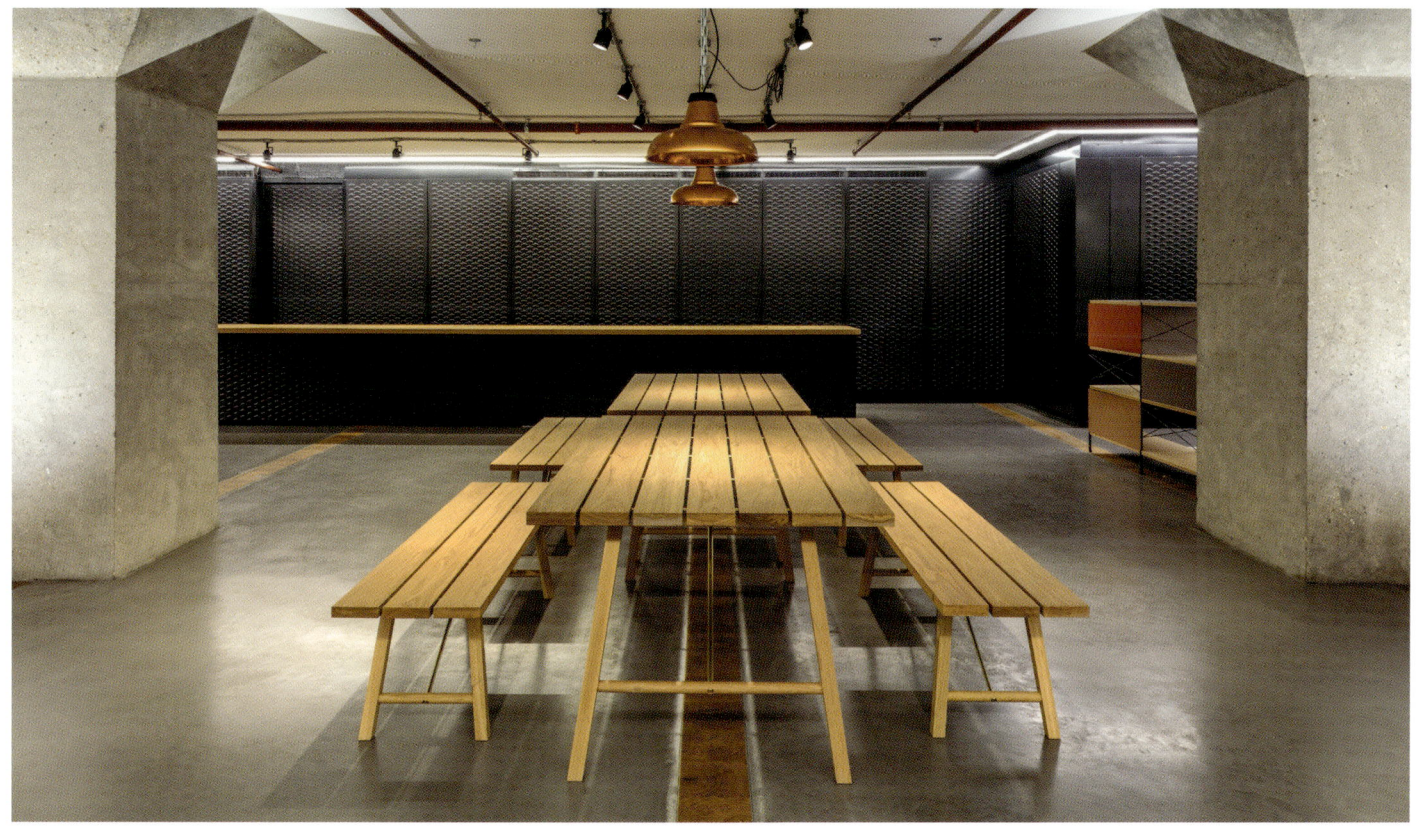

Figure 3.9

These central pendants provide ambient downlight over the table, with the track spots offering accent lighting. These are supplemented by the perimeter concealed linear source, which provides general light, ensuring the room edges are not dark.

Downlighting

Downlights are the most common method of providing ambient light. They can be surface- or track-mounted fixtures, cone-shaped or wide-beam recessed downlights. Light falls in a wide cone and provides an even illumination. It is a functional effect and allows light to fall directly onto the surface below. How close the light source is to the wall will determine how much of the wall is lit.

It is important that the light source is concealed. This is usually achieved by setting the light away from the surface of the luminaire within a deep baffle (25–40mm) to avoid the source being visible directly. Unfortunately, many budget fixtures expose the light source. The baffle is important to prevent reflection of light, so is often matt black. White baffles are less visible in the ceiling in daytime but white reflects light and so when the light is on, the baffle becomes almost as bright as the light source itself, negating the original purpose of the baffle. Our eye is drawn to the brightest point and if we see the light source, this becomes the focus, rather than the area or object that is intended to be lit. Accent lighting, described later, can then be used with a narrow-beam focus to cut through the general downlight effect.

Figure 3.10

Decorative Lighting

Decorative lighting may be pendants,
wall lights or table or standing lamps.
Often a combination is used, and the
style is determined by the architecture
and interior design. Many interiors will
use decorative fixtures for ambient
lighting. It is important they provide
sufficient light. Some modern pendants
and wall lights are more for effect as
artworks rather than a useful source
of light. Other sources of ambient light
will then be required.

ACCENT LIGHTING

Figure 3.11a
Figure 3.11b

These images show the power of creating a focus using accent light. The pottery on the landing is lost without light (a) but is dramatic when lit (b).

Accent lighting can bring a room to life by creating a focus of light that adds contrast or highlights a particular feature. It is like a beam of sunshine which captures a specific area or object, creating a play of light and shadow.

A shaft of sunlight breaking through clouds brings a sense of excitement and pleasure. Accent lighting achieves the same effect by focusing a narrow beam of light on a favourite object. A room lit purely by ambient light shows everything but features nothing. It will be bland. By concentrating on a particular object (a trick often used in retail) the eye is drawn to this focus, yet one still notices everything else around. Contrast makes things appear more interesting.

Selecting a narrow-beam light source with a high Colour Rendering Index (CRI – see Chapter 4) will ensure colours are more intense. The colour of objects appears to change from dawn to dusk as the colour temperature of the light changes through the course of the day.

Deploying the power of light to capture attention requires skill. Deciding what not to light is as important as choosing what to light.

Figure 3.12

A narrow beam provides a central focus on the flowers. This contrasts with the ambient light created by uplighting the textured wall.

Spotlighting

Spotlighting is achieved by using a narrow-beam directional downlight or a track-mounted spotlight. It's ideal for focusing light on an object or piece of art and should be controlled separately to the ambient light. In traditional houses, picture lights can be more appropriate than spotlights for lighting art.

Figure 3.13

The black double downlights over the kitchen island provide the task light. These were chosen as their contemporary feel reinforces the design of the kitchen. The ambient lighting is from the uplight above the tall kitchen cabinets and the reflected light from small directional downlights around the perimeter; these are baffled so almost seem to disappear. In the window space, miniature uplights emphasise the chamfered reveals. The combination of effects creates layered lighting; by controlling each element separately, different moods can be created.

Task Lighting

Task lighting is used when light is required for a function such as reading, writing or cooking. General light is usually not enough and extra brightness is needed.

CHAPTER 4

WORKING WITH LED

LEDs have now taken over from traditional tungsten, halogen, florescent and other sources as the main source of light for residential and commercial buildings. Understanding their strengths and limitations is important.

Figure 4.1

The use of concealed linear LED to float the folded ceiling contrasts with the multiple point sources used at pool level rather traditional larger underwater fixtures.

RETROFIT LEDS

This chapter reviews the best selection of LEDs. Many retrofit LEDs have been developed to replace halogen lamps. They vary in quality enormously and should be rigorously checked for colour temperature, quality and dimming.

UNDERSTANDING LEDS

Light-emitting diode (LED) is the name given to a semiconductor diode that emits light when conducting electric current. Originally used widely in electronic displays, LED is now used for outdoor and indoor lighting because of its light quality and efficiency. The best LEDs can be expensive. One needs to consider colour quality and consistency. LEDs have a long life: some last for 50,000 hours compared with halogen's 2,000 to 3,000 hours. Therefore, investing in quality is wise – if only to avoid living for 50,000 hours with a dull, flat light with poor colour quality.

KEY CRITERIA IN LED SELECTION

Three criteria to consider when selecting LEDs are colour temperature, colour quality and colour consistency. The lighting designer should also understand thermal properties of LEDs and optics.

Colour Temperature
This is a measure of the coolness or warmth of the light. It is measured in kelvin (K). To compare LEDs to natural light, at dawn sunlight may be warm at 2,200–2,400K; at midday it is whiter and cooler at 5,000K and by sunset may be 2,000–2,400K.

Unusually, the higher the kelvin, the cooler the light. A surgical theatre might be lit with 5,000–6,000K lights, an office 3,500–4,000K and a retail store 3,000–4,000K. In food stores, the fish display will use a cooler colour temperature light, at 3,500–4,000K and the meat and bakery departments may use a warmer colour temperature of 2,700K to enhance what they are selling.

For the home, a warmer source of around 2,700K is usually best and possibly a bit warmer with chandeliers and decorative lamps, at 2,200–2,500K. Halogen light was 3,000K at full power and when dimmed around 2,200K. The old-fashioned tungsten lamp was around 2,400K and dimmed to 2,000K and the exposed filament lamp is around 2,200K. Originally it was hard to get LEDs to be warm, but now technology has developed so much that almost anything is possible. Sometimes in hot countries, a cooler colour temperature of around 3,000K is preferred for daytime.

Figure 4.2

The colour temperature of natural light changes during the day.

The advantage with halogen and tungsten was that dimming the light source would affect the colour temperature, which became warmer. Originally LEDs, like fluorescents, would maintain their colour temperature when dimmed, which produced a flat, grey, dull light, particularly if the CRI (Colour Rendering Index, discussed later) was poor.

These days, colour-tuneable LEDs are available, which dim from cool to warm. These must be selected with care and the CRI should be checked, as often the light when dimmed can appear too yellow.

Figure 4.3

A cool 3,000K LED linear strip is used in the cove during the day.

Figure 4.4

A warmer 2,400K LED linear strip gives a softer evening effect.

Figure 4.5a
Figure 4.5b

The fruit bowl with a high CRI (a) makes the fruit look vivid and the colours jump out. The fruit bowl with the low CRI (b) seems flat and lifeless by comparison. The colours can be identified but seem dull.

Another issue is the difficulty of achieving a very narrow beam for accent lighting in miniature fixtures. The warm, dim light source is normally created using a combination of LED chips, maybe 2,200K to 3,500K. By combining the LED chips, either individually or as a COB (chip on board), a larger base is needed, which becomes difficult with miniature fixtures. If using tuneable fittings, select those that go as low as 2,200–2,400K. If the lowest is 2,700K (sometimes described as warm), this will be too cool at night and will not provide the desirable warm glow similar to the old tungsten lamp. A colour temperature of 2,700K will be fine during the day and early evening, but later in the evening the warmer light associated with candles and firelight is what we sometimes seek to create.

If not using tuneable LEDs, one solution that still allows a warm effect is to use 2,700K for architectural uplights and downlights that can be miniature, and 2,200–2,400K for decorative light sources such as wall lights, lamps, pendants and warm linear strips in coves and joinery. These will then provide your warm late-evening light.

If a linear strip is also used during the day for general light, then either use a second cooler linear strip beside the warm one or find a combined tuneable offering with a desirable range of 2,000–3,500K.

Colour Quality
This is sometimes referred to as colour rendering and is a vital consideration in choosing a correct LED. Numbers on the Colour Rendering Index (CRI) refer to a light source's ability to render colour accurately compared to a reference light source which best replicates daylight. It gives us information on the colour fidelity.

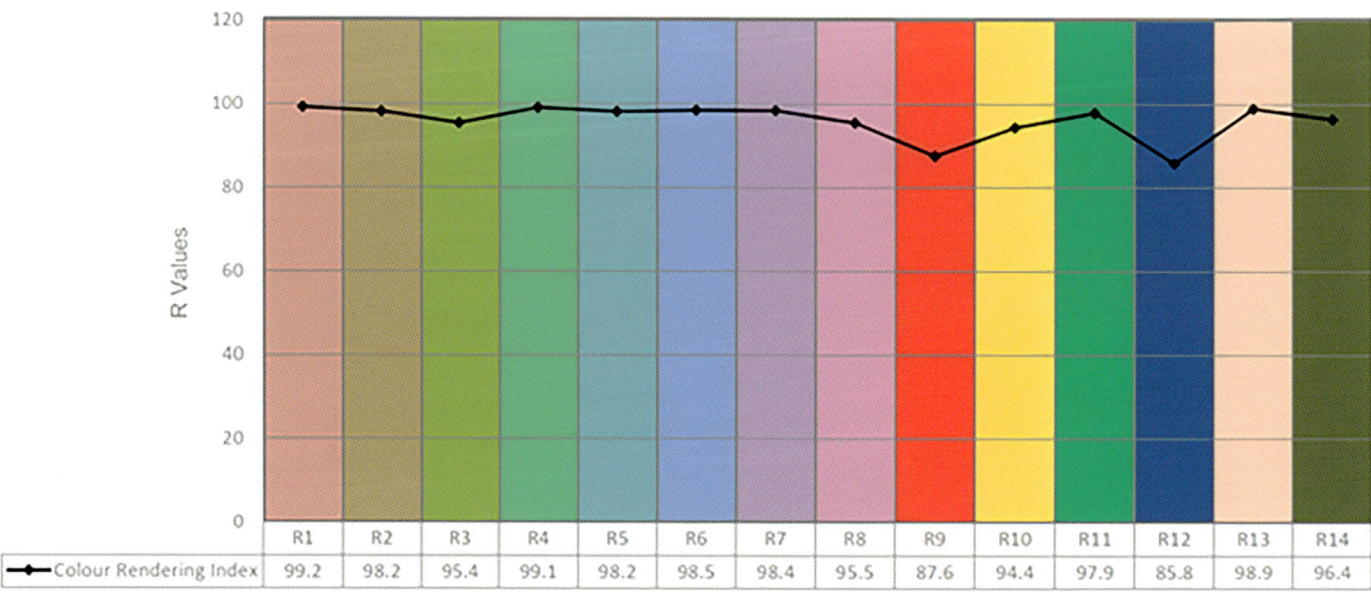

	R1	R2	R3	R4	R5	R6	R7	R8	R9	R10	R11	R12	R13	R14
—◆—Colour Rendering Index	99.2	98.2	95.4	99.1	98.2	98.5	98.4	95.5	87.6	94.4	97.9	85.8	98.9	96.4

Colour Rendering Index (CRI)

The CRI is a scale from 0 to 100% indicating how accurate a light source is at rendering colour when compared to a reference source such as sunlight. Often, manufacturers try to get better results by referencing only eight colours, with each colour being given a rating between 1 and 100, the average of which is stated as an Ra value. (Ra is the unit of the Colour Rendering Index, with which the quality of colour reproduction of lamps is described. The higher the Ra value is, the less the artificial light distorts the colour of objects. The maximum Ra value is 100.)

The CRI can also be referenced against R14 colours (as shown in Figure 4.6). Red is notoriously the hardest colour for LEDs to render as they naturally favour the blue spectrum. So, realising a high red will give an excellent visual appearance. This is important particularly for artworks. If the CRI is high, the quality of the light feels better and makes everything look more inviting. A good LED needs a 95 CRI as a minimum from the R14 range. High CRI LEDs are more expensive. Lower CRI LEDs should be used for storage spaces and locations where colour rendition and quality is less important.

Figure 4.6

The Colour Rendering Index (CRI). The plotted line on the graph shows colour rendering values across 14 sample colours.

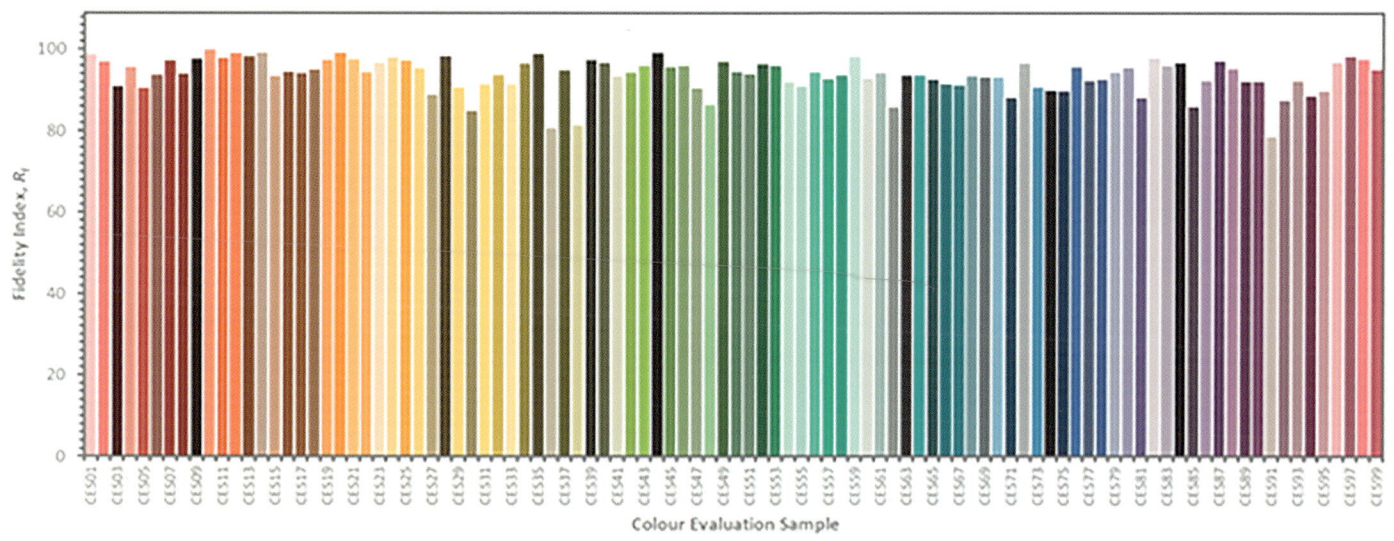

Figure 4.7

The TM-30 colour fidelity index.

TM-30

The TM-30 method is an alternative and more accurate measure of colour quality. Instead of using 8 or 14 colours to measure fidelity, this uses 99 sample colours. With more colour samples, a more accurate assessment can be made. The inspiration of the extended colour range has come from nature: fruit and vegetables, including oranges, red onions and green apples, and the new test colours have a similar high saturation to natural colours.

The old CRI index cannot be directly compared with TM-30 as CRI considers only the average accuracy of a limited colour sample and measures just the fidelity not the saturation of colour. The TM-30 widens the selection. The method offers two measures: colour fidelity (RF) and colour gammet (Rg). Colour fidelity is like CRI and references the average value of 100 swatches rather than 8 or 14. This larger selection matters as there are many more colours than 8 or 14 in the visible spectrum. Using such a limited sample to describe colour rendering could not capture an object's visual appearance under an artificial source of light compared to natural light. Colour gammet is the average level of saturation relative to a reference illuminate (measured between 60 and 140). Saturation is the colour's intensity versus white – a higher value details a better rating: values above 100 are visualised with increased saturation and values below 100 represent a decrease.

Colour Consistency – Binning

Colour consistency is an important issue to consider, and is also known as colour binning. It's the process of sorting LEDs by quality characteristics, such as colour, voltage and brightness. To measure and control colour consistency we use an index relating to the light sources' variation in chromaticity (colour). If not tightly controlled, this difference in colour can be picked up by the human eye and would be unacceptable. In selecting lights for an interior, you want them all to match and be selected from the same batch. Colour difference can be identified by steps of a MacAdam ellipse. The scale is set by the ANSI Chromaticity Standard.

An LED chip is made up of a small diode, created from the semiconductor material. When power passes through the diode, the movement of electrons through the chemical layers creates light. The diode will often create either a blue or indigo light. To convert this colour light to white, we coat the diode, or die, in a thin layer of phosphor, often yellow or orange in colour. The mix of phosphor, paired with the diode, dictates the light quality of the LED – including CCT (correlated colour temperature) and CRI.

A MacAdam ellipse can identify a simple colour on a point of a graph with an ellipse around that point. Within the ellipse there would be no discernible colour difference from the reference point at the centre. LEDs can have inconsistencies between two LED strips and hence modules built from them. In the early days, LEDs were sorted by separating them into various bins according to colour and output (hence the term 'colour binning'). A high-quality LED module will use LEDs appearing within a one or two MacAdam ellipse, where there will be little variation in perceivable colour.

Thermal Management

A common misconception is that LEDs do not generate heat. The beam of light is cool, but LEDs convert power into heat and the life of the LED chip is directly linked to its operating temperature. This is why LEDs are usually mounted on a heatsink that draws heat away from an LED chip (usually metal). Sometimes this can be the body of the light fixture itself. The heatsink regulates the temperature of the fixture.

If an LED operates above its thermal guidelines, not only can its life be reduced, but there can also be a colour shift visible in the LED. In hot regions such as the Middle East, it is important to know the ambient temperature to understand the lifetime and colour consistency of the LED. Product designers need to ensure the appropriate heatsink for the environment.

Optics

The next feature is understanding the optical control of the LED. Optics are specifically designed for different LEDs to ensure that the light can be focused. They offer varying beam widths: very narrow (6°–10°) is hard to achieve; more usual is narrow (12°–19°), medium-beam (24°–27°) and wide (36°–45°). Different manufacturers offer fixtures with different optics in their luminaires. Lighting designers select fixtures according to the light effect required. For example, if a very narrow beam is required to focus on a specific object in a room with a high ceiling of, say, 5m, the selection may be 6° or 8° to ensure that a tight beam of light is achieved. A medium or wide beam would be more glaring and also provide too much of a wash of light. That same optic on a lower ceiling height may be too narrow and concentrated to achieve the desired effect and a wider beam will be required. So, an understanding of trigonometry and of the concept of the angle of light and how it falls is needed. If more general light is required, then the beams of light must overlap rather than creating pools of light and shadow. These are all decisions the designer makes when choosing a light fixture.

COLOUR
AND LED

One advantage of LED has been that its miniaturisation has allowed different colour chips to be used together, either in a strip or fixture, to create colour change. This is best if combined with white. This is because the white colour created from RGB (red, green and blue) always appears grey; having a true white as part of the mix is desirable. Colour-changing lights have their place in hospitality, fun rooms and children's rooms. By controlling each RGB channel, one can create a multitude of colours. It is often the secondary colours that appeal – such as magenta (combining blue and red), cyan (combining blue and green) and yellow (combining red and green).

LIGHT
OUTPUT

Having established the quality of the LED and selection process, we now must understand how to achieve the maximum output. The same LED can be made to deliver more output depending on the heat management, determined by the size of the heatsink. If the heatsink is large enough, then the LED can make use of a driver with a high current and deliver more light. If a driver with too high a current is used, the LED will burn out.

The LED driver is a device that regulates the current and voltage, and reduces 230V mains to the amount required by the LED, commonly between 3 and 40V. It must provide sufficient current to light the LED at the required brightness and will keep this constant to prevent damaging the LED by overpowering (hence the name 'constant current driver' – see the next paragraph). (Similar to traditional transformers, LED drivers transform mains voltage AC to a lower DC voltage.) Many drivers are now programmable for extra flexibility. To select the driver and the number of LEDs it can run, divide the maximum output voltage of the driver by the forward voltage of your LEDs within the LED fittings. The LED driver will also need to be selected for the correct drive current of the chosen light fitting, such as 350mA, 500mA, 700mA. A driver will have a maximum wattage associated to it, such as 20 watts, so the fittings should not exceed the maximum wattage of the driver specified. The drivers should be as close to fittings as possible or, if located remotely, then distance will incur a volts drop, so this needs to be considered when specifying driver voltages, along with cable sizes.

There are two main types of driver. The first is a constant current driver, which varies the voltage across an electronic circuit to maintain a constant electrical current. This means if there is a fluctuation in voltage, the current driven to the LED will be maintained. These tend to be used for individual LED fixtures with a single light source, such as downlights and uplights. The other type is a constant voltage driver, which has a fixed voltage of 12V DC, 24V DC or 48V DC. These are used for LEDs that require one stable voltage and have a current that is regulated. These tend to power LED linear strips and track systems.

CHAPTER 5

CONTROLLING YOUR SCHEME

Controlling light is key to controlling mood. With multiple circuits and layers of light, it is possible to change the mood of a room to create drama and make a space work differently from morning to night. If set up properly with the right lighting design, any room can be transformed at the touch of a button from a bright, working place to an intimate party atmosphere. Lighting, more than any other design medium, can create flexibility in any space.

First, consider how many circuits or switch lines are necessary. Each lighting effect – offering different moods – must be independently controlled. For example, if you have a decorative pendant and downlights, each will dim and be controlled separately on a different circuit or switch line. This is true for decorative lights (e.g. lamps and pendants) and for architectural effects (downlights, uplights and LEDs). By controlling each effect separately, you can change the concentration of light, change the mood of a room and create changes throughout the day and night.

STYLES OF
CONTROL PLATES

Figure 5.1

Once the integrated shelf lighting is on, the edges of the room come to life.

Each manufacturer has its own range of controls and each country has different designs. It is possible to have custom control plates which can also be used to control blinds, windows and other devices on the same plate.

Standard Rotary
The simplest form of dimmer is the classic rotary dimmer, which can dim lights up and down from one location using a separate rotary knob for each circuit/switch line.

Momentary Dimming Control
This is a push-button dimmer with a remote dimmer pack. It allows lights to be dimmed from multiple locations. The action of pushing or holding a button sends a signal to the remote dimmer pack or driver to dim up/down the lights. These switches tend to be low voltage and can be located in bathrooms, unlike mains voltage switches or rotary dimmers.

Figure 5.2

Here, the lighting is programmed to be practical rather than mood orientated, mainly downlights over the island and table.

Figure 5.3

The mood of the kitchen is softened by indirect elements of light from under the island, the soft uplight behind the banquette, the focus on the backlit sculpture and the delineation around each skylight. Each effect adds its own layer of light.

Scene-Setting Systems Control Plates

The next form of control is preset. Instead of multiple buttons operating different lighting effects individually (e.g. pendants, picture lights, shelving lamps), the buttons recall a complete scene. This requires a remote dimming system, and each lighting effect is adjusted individually from an app and then set with its own unique settings for each circuit and memorised to create a scene. This allows the perfect mood to be recalled at the touch of a button. The style varies enormously, and bespoke plates can be used with most dimming controls. They can be designed to suit an interior's style. Customised plates using a retractive switch can be made compatible with most systems. It is best to arrange the scenes in the same order in each room of a house for ease of use. Avoid having too many scenes on a switch as this can lead to confusion. I usually program four scenes or fewer and allow for a raise/lower facility for each scene.

Suggestion of Scene Settings

Scenes could be programmed as follows:

Scene 1/button 1: Bright (used during the day)
Scene 2/button 2: Medium or soft (used for early evening)
Scene 3/button 3: Moody or low (used for dining)
Scene 4/button 4: Low or night (used for TV viewing)
Finally, off.

In each scene, the light settings will be adjusted and then saved. Typically, this is done from a program or app on a computer, tablet or phone. A technician was previously required to program settings but today it is easy to adjust and fine-tune each scene oneself after the technician has written the program.

Configuring the buttons in each room from bright to dim keeps settings consistent through the home and is easy to remember.

Suggested Scene Setting for Lower-Use Rooms

Scene 1: Bright
Scene 2: Soft/mood
Off.

Cloakrooms and utility rooms may need only two scenes. The key is simplicity.

If more than four scenes are required in any room or to link scenes in different rooms or the whole house, it is best to store these on a phone or tablet. An all-off function by the front door is useful. I would always do this by double-tap or a long-hold function programmed on the button so that it cannot be activated easily by someone who does not know the function. You want to avoid guests accidently plunging you into darkness.

When to Use Preset Scenes

I usually use preset scenes when there are more than three or four circuits within a room. Otherwise, it is a bore to keep adjusting four rotary knobs or push buttons to get the correct mood when it could be done instantly with the touch of a button. Preset systems work particularly well in open-plan spaces.

There are many systems available, from remote hard-wired systems to wireless. The selection will depend on the size of the property and budget. Often it is the selection of the switches that determines the choice of a system; some people prefer traditional rather than technical-looking switches. Today, there are plenty of third-party switches that can be used together with an interface of your selected control system, allowing more choice in the appearance of switches. Whatever the selection, it must be simple to use and labelled clearly.

Remote wireless control is useful for retrofit developments and listed and old homes where construction may limit wiring. It makes mounting control panels on a stone wall easier as no cables are required. The battery life of these switches is often five years.

Preset systems used to have a bad name because, if not set up properly, clients couldn't adjust systems themselves and needed to call an engineer. This is no longer the case.

Figure 5.4a
Figure 5.4b

An apartment entrance hall can be fine-tuned to appear welcoming during the day (a) and moody in the evenings (b).

Figure 5.5a
Figure 5.5b
Figure 5.5c

Bright (a), soft (b), low (c). In this bathroom, the brightness is created by lighting the tiled wall; if there is only one light over the bath and the wall is darker, it makes the whole space darker. At night, for a spa-like feel, you could introduce uplights behind the bath.

Each scene needs to be adjusted individually to achieve the appropriate balance. For example, on scene 3 (the lowest setting), picture lights may be at 70%, a pendant at 25%, shelf lights 40% and downlights 20%, whereas on scene 1 (the brightest), downlights could be at 100%, shelf lights 0% and pendants 80%. It is this careful balancing that makes all the difference.

DIMMING
PROTOCOL

Mains/Phase Dimming

This is a traditional method of dimming halogen/incandescent lamps with leading-edge dimmers by altering the voltage passing through the lamps. The more voltage reaching the lamps, the brighter they will be. With the introduction of LEDs with electronic drivers, more compatible methods of mains dimming were developed using trailing-edge dimmers. It is important to check the compatibility of traditional lamps and LED lamps with the mains dimmers as they sometimes have a minimum load requirement. Mains dimming was initially rarely used for LEDs as the mains dimmer meant the LED source did not dim very low and often flickered. However, this has been resolved with improved technology of both drivers and dimmers.

For some lights, the driver is integral to the light source and for others it is remote and the choice of dimmer must be compatible with the driver module.

1–10V/0–10V dimming

This is an analogue system and easy to install. It uses a mains cable to switch the power on/off; the addition of a dimming signal cable links the drivers to the dimmer, providing the signal to dim the driver. The signal wires should avoid sources of electrical interference. It is important to select 0–10V drivers (which dim to zero) rather than 1–10V drivers (which dim to 10%) as the lower dimming performance will be better, as at 10% an LED can still appear too bright.

DALI dimming (Digital Addressable Lighting Interface)

This is often the preferred dimming method for architectural LEDs and is considered smart dimming. It is a protocol for sending a digital message along two conductors to control the luminaires. The signal cables link the DALI drivers in one loop, unlike mains and 0–10V dimming when the circuits are already individually wired as separate channels. Being linked in a single group, less cabling is required. The drivers can then be addressed individually to regroup them to make the required circuits. This allows for total flexibility and changes of control, even at the end of the project, as each driver is addressed and can be assigned a group, so is easily changeable by readdressing. DALI has a standardised dimming curve. It dims very low to 0.1% using the correct driver. A single DALI network is limited to 64 devices and these include drivers, scene plates, PIR sensors, etc. Using an advanced system with multiple devices will require expert commissioning, although once done the fine-tuning by the owner and designer is possible. DALI can also be wired in the traditional manner as predetermined channels, like mains and 0–10V described earlier. This is known as a broadcast and will not require the same level of commissioning.

DALI drivers contain a unique serial code which allows them to be uniquely addressed by the software and controlled individually to different levels, if required, when wired onto the same DALI cable. However, when DALI broadcast commands are sent to the drivers, all of them on the same cable will go to the same levels, which makes this a less flexible option.

Bluetooth Low-Energy Control (BLE)

Bluetooth low-energy control operates at 2.4 GHz and is the same frequency as, for example, Zigbee and Bluetooth classic. It is ideal for the control of lights between luminaires and a switch sensor. A reliable system, it is already built into smartphones so can easily be used to control lights remotely, using an app from the relevant manufacturer. The cost of cabling is reduced as the BLE control creates a mesh to communicate a signal between luminaires; each has a remote BLE module to control 0–10V, DALI or phase (mains) fixtures. Some luminaires embed the module within the fixture, so communication can be made without the need for physical cables, ideal, for instance, when mounting a switch on a stone wall. Careful consideration is required to ensure good wireless coverage and your manufacturer can guide you on this.

The dimming curve adopted tends to be the DALI curve, which is becoming standard and avoids the variety of dimming curves and 0–10V analogue systems. The system is flexible and is being used more widely.

Time Clocks and PIRs

Other forms of control include time clocks, where the lights are set to turn on/off at preset times. A PIR (passive infrared presence detector), activated by the heat of a body, is frequently used for after-dark security lights or wall lights and even with garden lighting schemes on the approach to front or back doors. It is often used to bring on a night-light in a bathroom or to bring lights on automatically in storage areas, etc. The PIR has a timer to set the duration it should remain on and can also have a daylight sensor to prevent daytime operation when used externally.

The infrared and time clock systems can also be used as part of a lighting control system, discussed earlier, where each of the lights can be programmed to integrate with an astrological time clock to automatically adjust for seasonal changes. The programming instruction is usually linked to sunrise or sunset, bringing lights on half an hour before sunset, for example (so there's no need to change the time clock for seasonal changes). Many manufacturers have developed Wi-Fi or Bluetooth versions of their controls that can be used separately or linked to the main system. This is ideal for adding a control circuit without changing the whole control system.

In addition, PIRs on control systems can be used to trigger a scene of multiple circuits. This feature can be used in conjunction with programmable logic, so the PIR will trigger different scenes at different times of day or not trigger on the outside lights during daylight hours.

The next development in lighting controls is incorporating sensors into light fittings themselves. They will only work in residential properties if there are solutions for the decorative fixtures as well as the architectural lights, as it is the combination of both effects that creates the magic.

CHAPTER 6

HOW TO CREATE A LIGHTING PLAN

In previous chapters we have discussed the various elements of light and the effects that can be achieved; now we need to demonstrate the best way to translate design ideas into a plan so the installer can implement the scheme successfully. In this chapter, I show how two-dimensional lighting plans translate into fully lit, three-dimensional reality, using a contemporary pool-house as an example.

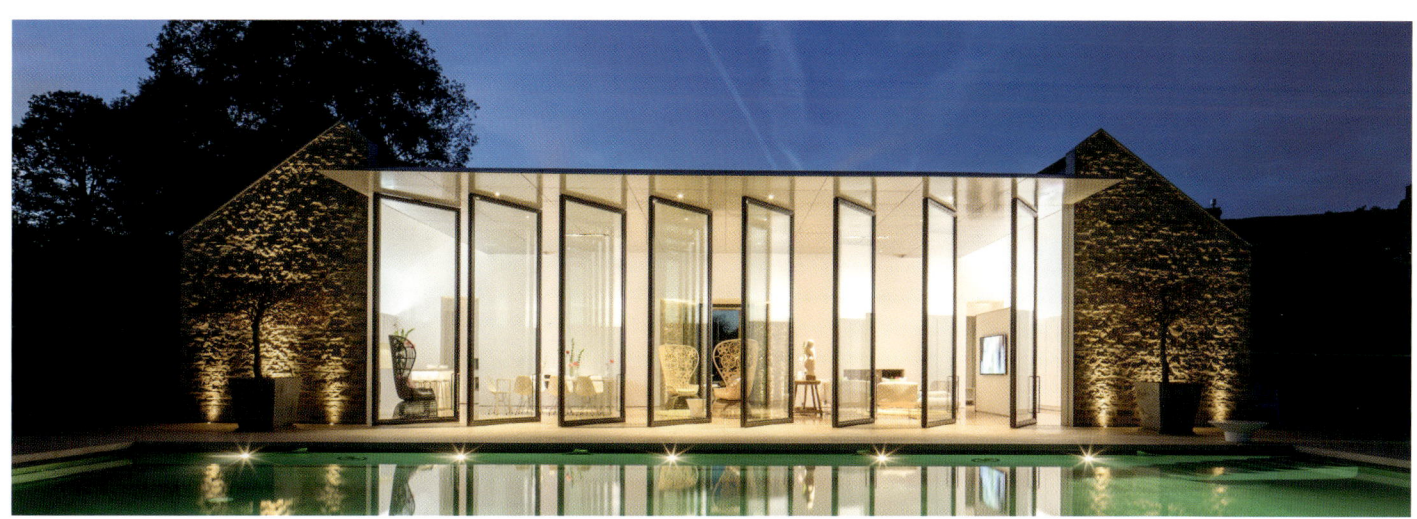

Figure 6.1

The exterior vision of this project contrasts an old textured wall, emphasised by uplights, with full-height contemporary glazing, leading out to a pool. The open-plan interior has numerous light settings for different moods, all marked clearly on the lighting plan.

WRITING THE LIGHTING PLAN

In this section, we review what's important for a lighting plan. The illustrations and photographs use the plans and final images of one of my projects, the pool-house pictured in Figure 6.1.

The value of the lighting designer is understanding light in three dimensions, and how it affects every surface in a space. Too often, lighting is an afterthought on an architectural plan, considered only in two dimensions, with elevations largely ignored. But when we enter a room, we view the whole space – often looking to the walls first. So the lighting plan must relate to walls and every architectural detail in each corner of the room, as well as to the furniture and fixture layouts.

When preparing a lighting scheme, you need details of the space, ceiling heights, elevations and the arrangement of furniture. The lack of this information is a reason why so many schemes have a boring grid of ceiling lights installed to provide a general light level but with no consideration of the space or how the room is to be used. These grids are often selected as they look symmetrical on the ceiling plan, which may be fine if the room is rectangular, as in this pool-house study, and arranged symmetrically. However, they take no account of the space below where the light will fall, and are often used when a generic layout is put on the plan.

Creating a successful lighting plan involves building up layers of effects and then translating them to the two-dimensional plans. This involves clear communication with the client, architect and interior designer. The more information we have, the better the lighting scheme will be.

KEY

L1	DOWNLIGHT	L7		RGBW COLOUR CHANGE LINEAR	
L2	IP RATED DOWNLIGHT	5A		5AMP FLOOR SOCKET	
L3	SQUARE DOUBLE DOWNLIGHT	5A		5AMP WALL SOCKET	
L4	UNDER CABINET LIGHT			HIGH LEVEL (ON/IN CEILING)	
L5	RECESSED UPLIGHT			OTHER	
L6	2700K LED LINEAR				

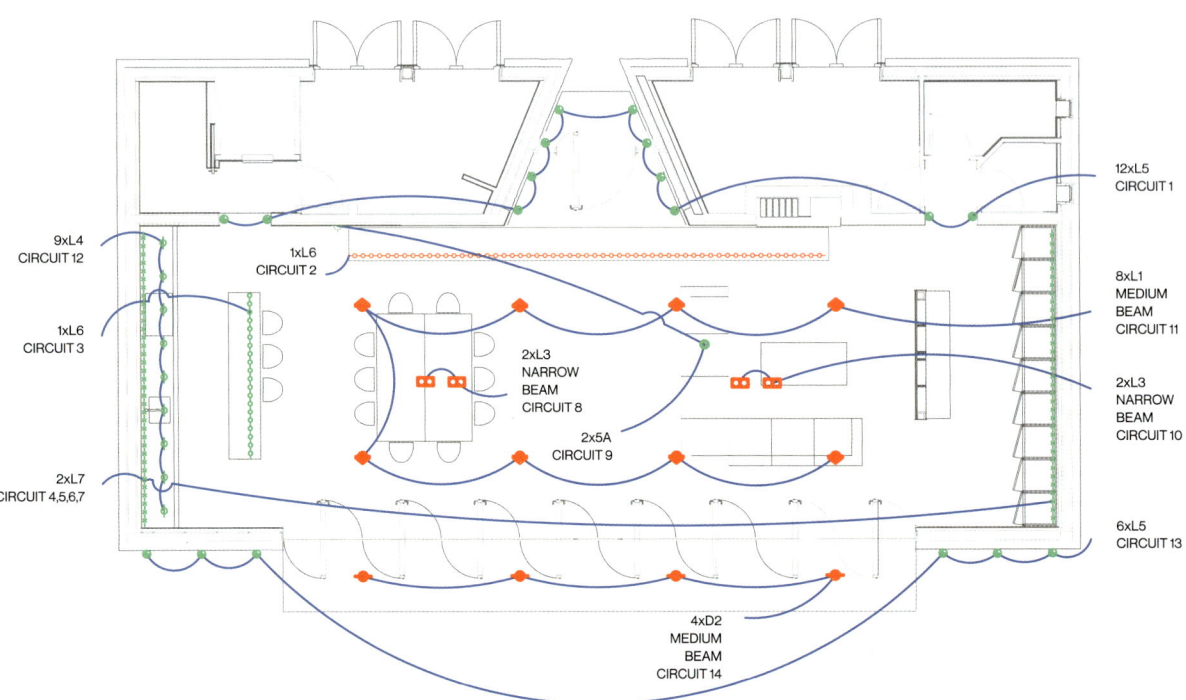

Establish a Key of Lighting Symbols

First, a clear key of lighting symbols must be established, ideally using individual symbols for different elements, such as downlight, uplight, socket, wall light, pendant, LED strip, shelf light, etc. There may be several symbols for different sizes of fixtures. For precise identification they could be given a fixture reference, which would refer to a specification. There is no standard. What is essential is that the key is clear on the plan and easy to follow.

Levels of Lighting

The next consideration on the plan is to differentiate between high- and low-level lighting. Sometimes there will be two different plans, with the ceiling lights on the reflected ceiling plan and the floor and wall lights on the floor plan. This helps the installer. All lighting can be shown on one plan, ideally linking back to a clear key. If colour-coded, the ceiling lights can be in one colour, with low-level lights in another colour, for added clarity.

The next element to consider is the lighting control and how this is defined. Sometimes circuit lines are used which link each of the lights to be controlled together and linked back to a dimmer switch. Sometimes circuit references are used to show how lights are linked back to a control system (e.g. C1, C2 are written on the plan and describe which fittings are linked together). These then refer back to a control schedule.

Figure 6.2

Using different colours for high- and low-level lighting, together with labels for fixtures, adds clarity to a plan. This lighting plan corresponds with the chapter's opening image Figure 6.1 of the glazed extension.

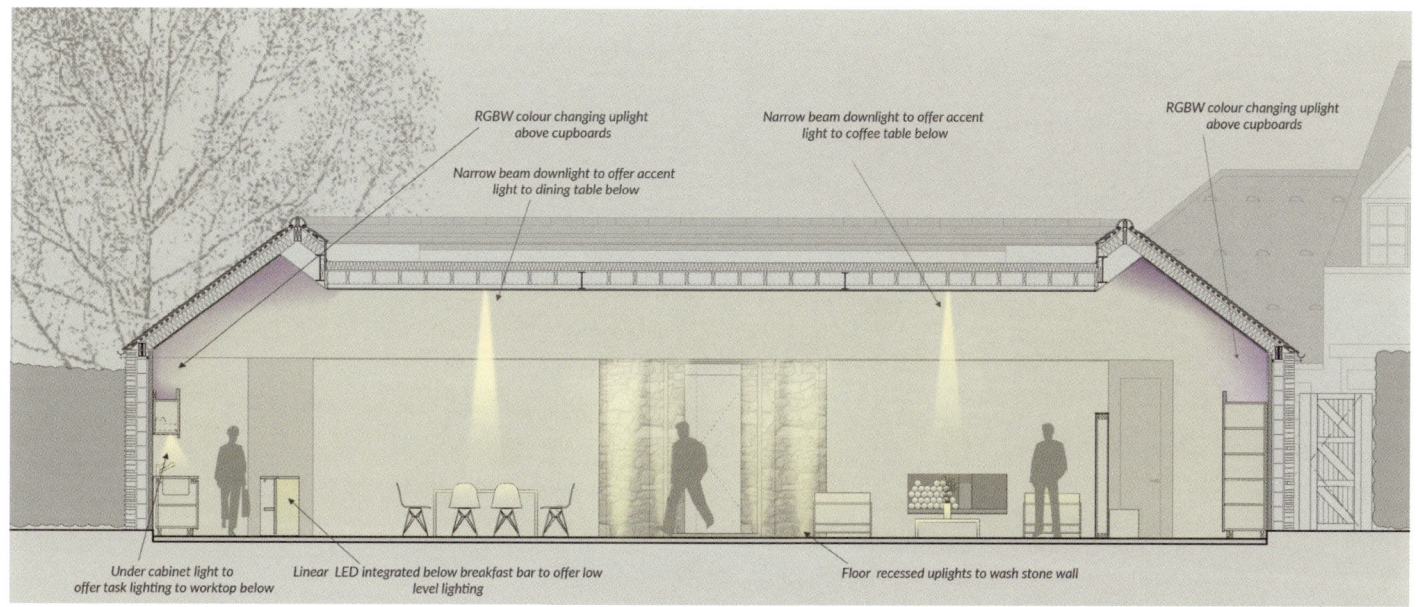

RGBW colour changing uplight
above cupboards

Narrow beam downlight to offer accent
light to coffee table below

RGBW colour changing uplight
above cupboards

Narrow beam downlight to offer accent
light to dining table below

Under cabinet light to
offer task lighting to worktop below

Linear LED integrated below breakfast bar to offer low
level lighting

Floor recessed uplights to wash stone wall

Figure 6.3

This section shows the levels where layered lighting should be installed and a diagrammatic effect of the lighting.

The key on any plan is clarity for the installer. Referring to elevations and specific details is important to leave nothing to chance. These details may be shown at the edge of the plan or in a specification document.

Sections, Details and the Specification Document

Showing lighting on a section or elevation makes it easier for the installer to understand the scheme and where lights are located. Details show the integration of the light source into the architecture. (A typical detail is shown in the bottom-right corner of the plan in Figure 6.2.)

The specification document is important as it provides full details of the luminaire. It will help the installer understand if a fixture has an individual driver or a grouped one to power a series of lights. If it is a grouped driver, the installer will need to find a location for the driver within a minimum distance of the lights. The specification document will inform the installer of the wiring and dimming protocols and often refer back to a control schedule. It is the combination of plan, specification and control schedule that ensures a full understanding of the scheme, and no errors.

The lighting plan may seem like the start of a project – but in fact it comes after having discussed and agreed what the client wants and what works best with the architect's and owner's vision. It's a key stage to get right. Then it's on to the excitement of seeing 2D become reality, with all the drama that creating new spaces provides.

THE FINAL
RESULT

Figure 6.4

The correct levels for each lighting fixture must be clearly marked on the plan for the final effect to be successfully achieved.

Figure 6.5

Downlights in the external canopy ensure the eye is drawn out to the pool beyond. Key elements on the console are lit.

The juxtaposition of old and new is what makes this pool-house project interesting. Understanding how uplighting the rough stone would look was essential when positioning the lighting, and was specified on the lighting plans. The contrast with the huge, glazed panels and volume of the space offers lighting opportunities at different levels. Through a narrow opening we are welcomed into a large room with a view to the pool.

Figure 6.6

This dramatic, contemporary open-plan room needs different lighting to delineate each area; these different elements are all clearly marked on the lighting plan, from fixture to circuiting and controls, as well as every light level. In this image, uplight above the cabinets lights the ceiling giving a soft ambience; focused light is on the table.

This open-plan room, with glazed doors out to the pool, has a kitchen, dining and seating area. A key element of the lighting design was how to manage the pitched ceiling, which could have been overlooked on a flat plan. Our solution was to uplight it from below so that soft ambient light reflects off the sloped ceiling. To create a more intimate atmosphere, small LED under-counter lights are used below the cabinets to provide a regular array of scallops. As the work surface is highly polished, individual sources were a better option than the reflection of a linear strip. A linear LED strip was used under the kitchen island to reflect off the white stools. Knowing the materials to be used, including the colour and texture of each material, affects lighting decisions, and must be considered early in the lighting design.

To achieve narrow pools of light over a 4m-high space we selected 8° very narrow beams, to focus on the table and coffee table in the seating area, with medium-beam downlights for infill lighting. It was crucial to have external downlights in the projecting canopy to draw the eye out to the pool. The correct position of these is indicated on the plans, to achieve the best overall effect. Lighting outside the windows avoids the effect of the glazing appearing as a mirror at night. The immediate lighting outside must be brighter than the inside.

Figure 6.7

A narrow focus of light on the central area and coffee tables creates the feel of a room within a room. A linear LED within the high-level skylight reflects off the wall, adding interest after dark. Knowing the position of furniture and the height of ceilings and position of skylights all impacts the lighting decisions made at the planning stage.

Figure 6.8a
Figure 6.8b

We enabled the light colour to change with linear RGBW LEDs, so that this space can switch to party mode. Clearly marking this on the lighting plans helps the lighting engineer install them correctly.

As well as the ambient light reflected off the pitched ceiling from both ends, above the kitchen units and full-height cupboards, we introduced colour using RGBW (red, green, blue and white) LEDs, helping to transform the room for parties.

PART 2

LIGHTIN
THE HO

GME

CHAPTER 7

FRONT DOORS AND FIRST IMPRESSIONS

First impressions count. The point of arrival sets the scene for the rest of the house, so it needs to feel welcoming. You may prefer an entrance to be discreet or to make an impressively grand statement. Some choices may be defined by local restrictions in planning: a listed terrace may require all entrances to appear the same; or in the countryside, lighting may be restricted to maintain dark skies and protect wildlife (such as bats, whose flight paths can be altered by bright light). The style of a house and its location affect how its entrance is lit. A stand-alone house with no planning restrictions may suit bold, original lighting, while a traditional home may suit a more understated approach.

Figure 7.1

This facade has been lit at each level to create drama and impact, with its large trees and sculpture lit. Small trees are backlit and the light catches the cornice detail below the balustrade.

THE FRONT DOOR

Figure 7.2

Wall lights are the visual focus here and appear to do the work. If this were true, however, the door would be in darkness, as it is set back by a metre. In fact, six miniature 1-watt uplights graze up the brickwork, lighting the white soffit. Spiked uplights behind the hedge soften the wall lights with more ambient light.

Figure 7.3

Soft uplights to the facade of this home are placed within the planting. Wall lights or a lantern would conflict with the architectural detail around the door. The solution is to highlight the pilasters either side and the arch above with miniature 1-watt uplights and a miniature downlight recessed above the door.

A traditional solution to lighting an entrance is a lantern in a porch or a pair of wall lights framing the door. Both use a single light source, which can be glaring, while everything else is in darkness. When dimmed, these lights give a warm, welcoming feel but, of course, provide less light. Concealed recessed uplights or spiked fixtures uplighting the facade from flower beds work well. They provide a soft ambient light and allow you to dim the decorative lights and achieve a perfect balance. This is a visual trick: the eye sees the decorative lights, which gives the impression they're doing the work of lighting the entrance, when it is really the hidden architectural light sources.

Figure 7.4

The entrance to this property is framed by two dramatically oversized pots, lit using narrow 10° 1-watt uplights recessed at their base. In addition, two spiked fittings in each pot uplight the trees.

THE FACADE

Figure 7.5

This terraced house shares a pillar with its neighbour, so wall lights are difficult to place and to have just one may feel unbalanced. The solution is 1-watt Lucca wall light uplights to the door and French windows, combined with discreet floor washers to guide visitors up the path. This successfully identifies the house in the terrace without making a big statement on the street.

Several options are available for making a statement on first impression. Lighting architectural details, such as columns or the facade itself, is one. Lighting entrance gates and the drive up to the entrance is another (particularly effective on a tree-lined drive). To light a facade, recessed low-glare LED uplights are a good solution. Using 1-watt fixtures with narrow beams, positioned close to columns, door architraves a higher output for the facade. Different lenses can be applied to change the distribution of the light, such as spreader lenses which widen and elongate a narrow beam so more of the wall can be lit. For contemporary houses, a recessed linear wall grazer may be best. This uses a linear fixture with multiple sources, 50mm apart, and spreader lenses to provide an even, linear wash.

With planting by the facade, spiked lights in the foliage will be concealed, an effective solution as uplighting the walls creates ambience and distracts from the glare of a wall light on its own. If you can, avoid glare from security lights which light the person rather than the house and, leaving the building in darkness, can feel threatening.

Figure 7.6

A line of uplit silver birch trees frames the approach to this house. The double-height reception reflects in the water and the pale brick is softly uplit from the flower beds with spike fixtures concealed in the planting. In winter, the sculptural effect of the trees is strong, allowing more of the building to be revealed. In summer, the uplighting focuses more on the foliage.

An inner-city building may have more surrounding light and may require brighter lighting to create contrast. Choose low-glare uplights with the light source recessed within the body of the luminaire.

If there are several steps up to the entrance, it is advisable to light these for safety. Miniature LED fixtures set into each side is a good solution or lighting each or every other step. Square fixtures provide a softer, wider spread of light, and the light source is usually concealed. Circular fittings can have a narrow lens to project the light more dramatically across the steps. Care must be taken as the light source may be visible from a distance.

CONTROLLING
OUTSIDE LIGHTS

Figure 7.7

By day, the rustic stone of this house is flattened in sunlight. At night, the close-offset uplights accentuate the texture of the stone and guide one between the two buildings to the entrance.

For automatic control, outside lighting can be operated using a movement or presence detector known as a PIR (passive infrared sensor or daylight sensor) so it operates only in darkness and switches on as someone approaches/leaves. You can use an astrological time clock to set different levels from dusk to dawn. The timer can automatically adjust to seasons. Alternatively, an all-night setting can be part of a scene-setting system (described in Chapter 5). The control is an override to an automatic setting, so the lights can be left on as a welcome to guests and when switched off will revert to an automatic setting. In all cases, it is worth linking exterior lighting to a light sensor to avoid lights being left on by accident during the day.

ENTRANCE
HALLS

Figure 7.8

*Graphic lines highlight the symmetry
and define the entrance into the main
reception room.*

Lighting the entrance hall needs careful attention as it sets the welcome and tone of a house.

The hall may be a narrow corridor leading from the front door to the stairs or, in larger houses, almost a room, with a table that may double-up as a dining room. Avoid a row of downlights down the centre of a narrow hall. These give a dull, uninspiring effect, are unpleasant to walk under, give an arc of light halfway down the wall and won't emphasise pictures.

A good approach is to combine focused downlights with other effects. These could be decorative wall lights and pendants which complement the style of interior, or more architectural effects.

In tall corridors and narrow entrances, pendants can appear to lower the ceiling.

Figure 7.9

This double-height entrance hall's focus is several large pendants hung at different levels. They offer general light and connect with the galleried area. Accent light on the central sculpture and artwork is achieved with discreet spots, controlled separately. In the distance, an illuminated courtyard extends the space by day and night.

Sometimes, several overscaled pendants can bring fun and originality to an entrance. Playing with scale can be exciting.

In corridors displaying art, a track with spotlights directed both ways is a flexible solution, ideal for a contemporary space with a concrete soffit. Alternatively, use recessed downlights directed both ways. Wall wash lights in both directions make a corridor seem wider, or add interest by taking an asymmetrical approach, for example uplighting the wall on one side and lighting art on another. This is particularly effective if the wall has texture. A coffer will introduce ambient light into a hall, providing diffused general light, an alternative to using a decorative pendant.

Figure 7.10

Here the long curtains are lit with square, narrow-beam downlights. A track on both sides allows for a flexible solution to a changing layout and art collection.

Figure 7.11

The art is the focus in this hall. The picture above the console is lit with a medium-beam recessed fixture (out of view). The picture under the sloping stairs is more challenging to light. The solution was a small surface-mounted spotlight at the edge of the sloping underside of the stairs.

If there is a central hall table with flowers below a pendant, adding two low-glare, recessed narrow-beam directional downlights brings focus to the table. If recessing is impossible, add two small surface spotlights, with a narrow-beam distribution to create focus on the table. These spots should be controlled separately to the pendant. If there is no central table, then focusing on art and sculpture can lift the space.

Figure 7.12

Linear LEDs are concealed either side of a polished Barrisol ceiling (a stretched PVC ceiling which can allow light through). In this situation it reflects what is happening in the corridor, making the volume seem greater. The stairs are lit from the inside only, to light the riser.

For a clean, contemporary effect, a floating ceiling effect using an LED perimeter glow for ambient light works well. Combine this with occasional spotlights on features and artwork. Interest can be created along a corridor by uplighting elements that help break the length of a corridor. Lighting either with small uplights or graphic lines can be effective to introduce an alternate focus and layer, using light and shadow instead of uniform lighting.

A house's entrance and hall must be well lit as, together with the staircase (which we focus on in the next chapter), they form the spine of the building. As rooms lead off the hall, it is essential to include dimming controls to create changing moods which are in tune with the atmosphere of adjoining rooms.

CHAPTER 8

STAIRCASES

Staircases sometimes don't get the design focus they deserve. They can offer an exciting opportunity, altering the energy of a home. You can play with effects that may be too much for the main living rooms but are perfect for transient spaces.

Linking transitioning zones helps set the scene between the rooms. Stairs may form the central feature of a building and can be sculptural, freestanding, architectural, cantilevered stone, timber or metal. Lighting them can be challenging as they must be lit for effect as well as safety.

Figure 8.2a

A square floor washer lights this stair. When combined with an uplight to the window, it emphasises the angled reveal (to let in more daylight).

Staircases can be a wonderful sculptural element as well as the link between floors. As an architectural feature, they have gone in and out of fashion. In many mid-20th-century homes, they were pushed to the sides or backs of buildings and ignored as an opportunity for stylish design. Recently, they have become a design feature again. The variety of contemporary staircases allows imaginative lighting opportunities.

In homes where staircases are narrow, to maximise the size of rooms, integrating lighting into the stairs can be an ideal solution to widen their appearance and create interest rather than just a utilitarian necessity. They may be too narrow to use a wall light and installing downlights into a sloping soffit is not ideal as the light would shine into one's eyes.

Figure 8.2b

Here, we chose the unusual solution of uplights at the edge of the stairway. Because the stair riser is open, lighting across each step would cause glare at the other side. By uplighting the wall, reflected light illuminates the step and, with half-glare shields on the fixtures, side glare is reduced. The uplights also highlight the texture on this rustic wall.

A simpler, more traditional solution is lighting just the landings and half landings. It may be appropriate to use a different approach for the stairs leading to the lower ground floor, as these areas are often less formal and narrower.

Clever lighting of the staircase can be a way to make it a central feature.

TECHNIQUES FOR LIGHTING STAIRS

Figure 8.3

A dramatic play of light and shade is created by lighting every other step with a small square light. The overall effect seems to lead upwards towards the head of the stairs.

Individual Stair Lights

Sometimes the effect of individual stair lights is subtle and sometimes exaggerated. One small, square floor washer can light each tread or every other tread. If the steps can be seen head-on, it is important to choose a fixture where the light source is concealed and the light is directed towards the step. Individual 1-watt stair lights work particularly successfully in narrow staircases where wall lights would be impractical. If the stairs are within two walls, round 1-watt stair lights that shine across the treads are possible: there is no glare from the side which is enclosed, and the space benefits from reflected light off the opposite wall.

Figure 8.4

The linear light makes the stairs appear to float off the wall. The reflected light helps illuminate the treads.

Linear Lighting Solutions

These days, linear solutions are easily integrated into treads. Linear LED lights can downlight or uplight the riser or create an illusion of separating the stairs from the wall, giving a floating effect. Linear lights can be concealed within a handrail. Concealing the light source is essential, otherwise the light source becomes the focus rather than the stairs. If light sources are exposed, there must be a clear design intent and glare must be considered. Linear LED has become more flexible and can be used exposed, if diffused.

When lighting is recessed under each tread, use an LED strip with an opal diffuser (an opalescent coating to diffuse the light). The advantage of a diffuse source is softer shadows. Avoid direct views of an LED strip without diffusers, because the multiple small dots of an LED can look cheap and create unwanted shadows and reflections. With open-tread stairs, lighting each of the steps individually will light both the step and the area below, extending the feeling of space below.

Figure 8.5

Linear LED lighting is recessed under each floating tread. This lights the tread and also the area below, which on a closed-riser stair can be left dark.

Another technique using a linear LED is to emphasise the rear wall of a staircase, where the light washes from each side, possibly onto a piece of art or a textured wall. The eye is drawn to the rear wall feature and the reflected light will light the treads. A linear uplight could also be used under a staircase to reflect under the stairs. Both solutions of lighting the rear wall will make a space seem larger as the eye focuses on the wall behind the stairs.

Figure 8.6

The brick wall is emphasised by a vertical LED linear grazer unit. This is an LED strip in a profile with a 10° lens (grazing optic). The strips are located either side of the stairs, with the light 10–20mm away from the wall. The strips are concealed on either side by a baffle at least twice the size of the LED fixture to ensure the light fixture itself cannot be seen and it is just the effect that is visible.

Figure 8.7

A globe pendant suspended through the centre of the stairwell lights the treads and is an alternative to having a lantern on each landing.

Decorative Solutions

A lantern at each landing is often enough if combined with wall lights on the stairs. For years this has been the go-to lighting solution for landings and stairwells.

With an open stairwell, it is often difficult to bring lighting nearer to the stairs themselves. An alternative to a lantern on each half-landing is to suspend a series of globe lanterns in the central void, each linking to the next one. This can be combined with another layer of light, such as picture lights or wall lights at the half-landings and landings. The hanging-globe solution has the advantage of being central, so offers more light to the stairs than landing lights alone.

Another successful technique, particularly useful if the stair void changes size between floors, is to suspend two or three individual lanterns at different heights through the stairwell. This has the same impact as hanging globes.

Picture lights are rarely effective when viewed from the side. But if used on an end wall opposite the landing, they can provide a focus from the landing as one goes up and down the staircase. It is a useful way to light art when recessed downlights are impossible because of the slope of the stairs above.

Figure 8.8

A contemporary light suspended through the stairwell has the same effect as traditional lanterns. It is backed up by a concealed linear detail that gives a soft wash of light down the wall, accentuating the curve.

The solution for lighting both modern and traditional staircases is often the same, but the execution may be different. If the ceiling is flat and there is no skylight, then a large pendant is an excellent solution to fill and light the void. A multi-hang of smaller pendants of varying sizes may be more cost effective than a large pendant and more fun. If these are ribbed or in metal mesh designs, shadows will decorate the surrounding walls with patterns.

LIGHTING SKYLIGHTS
ABOVE STAIRS

Figure 8.9

This staircase has a skylight at the top embedded with LEDs to give a starlight effect. A glazed landing ensures natural light floods the floor below during the day. By night, the glass is edge-lit with an LED strip concealed on each side in the metal edging to make it glow. The illuminated floor adds light to the floor below. The glazed wall to the living area ensures that the staircase is a strong visual part of the room, helping the space to feel larger and more open.

Sometimes skylights are designed at the top of a staircase to introduce natural light into the centre of a house. These are wonderful by day, but at night become a black hole. Creative lighting is the solution. One idea is a soft band of light around the skylight, using a linear LED source. For this detail, use either an indirect LED built into a slot or a graphic line effect with a trimless LED set into the upstand. This should have an opal diffuser. The reflection in the skylight can be interesting.

Another solution for a skylight is to use glass with small LEDs integrated within it, which appear as stars when a current is passed through the glass. During the day the LEDs appear as dots, allowing daylight to flood in. This is specialist glass and is usually a secondary layer to the standard glass of the skylight.

LIGHTING THE UNDERSIDE OF STAIRS

The underside of a staircase should not be forgotten. In some instances, the area is boxed off for storage and offers no lighting opportunities. If it is left open, remember to bring light into this area, as it will help to create a feeling of space. Simple solutions vary from just an uplight behind a log basket to a lamp on a table.

A contemporary solution is to uplight the wall under the stairs with a linear LED recessed into the floor to light up the underside of the staircase. Individual small recessed uplights could also be used to bring life to an area usually forgotten. These solutions draw the eye to the wall and can make the hall seem larger.

Figure 8.10

Lighting the underside of this staircase was achieved simply using 5-amp sources located behind log baskets; plug-in globe lights provide this glow.

Figure 8.11

This metal staircase with open treads is lit by the integrated light within the handrail, which not only lights the treads but also filters light softly to the floor below.

Lighting may also be incorporated into handrails or the skirting. Ensuring the source is concealed is always important.

CHAPTER 9

LIVING ROOMS

Living rooms are where we relax with friends and family, enjoy reading, watch television, listen to music and entertain. They are multifunctional – used for different leisure activities at different times of day – so lighting needs to be flexible, and the control system must be easy to operate so the lighting can adjust to suit changing moods. The rules of ambient, feature and task light apply to living rooms.

Figure 9.1

This contemporary living area has a skylight that introduces daylight during the day and at night glows with an LED strip set into the upstand. Also concealed within the skylight are spotlights to light the large art piece. In the foreground, narrow-beam 10° spotlights focus on the flowers and both coffee table and side table.

FOCUS ON FIREPLACES

Figure 9.2

General light comes from a pendant (just in view), while soft, middle-layer lighting comes from the shelving; a directional spot onto the mirror and miniature fireplace uplights complete the scene.

A fireplace can be the focus of a seating area. A fireplace with surrounds can look good with uplighting to emphasise detailing and draw the attention, even if there is no fire in the grate.

Contemporary fireplaces may have raised hearths that can be underlit and texture can be emphasised to create a visual focus.

Figure 9.3

This contemporary fireplace has a floating hearth, emphasised by the LED strip below. Above the fireplace, there is soft focus on the bronze material and stronger focus on the log store. The lit coffee table draws in the seating area and the metallic screens lit on both sides provide a visual divide between the living and dining areas. The chandelier provides a central focus and ambient light.

Figure 9.4

In this unusual space, uplights each side of the chimneypiece emphasise the metal cladding, while, to the left, arch doorways are uplit. A downlight on the coffee-table flowers and another to the central metal screen around the fireplace attracts our attention.

AMBIENT LIGHTING
IN THE LIVING ROOM

Decorative lighting can be an important part of interior decoration and a source of ambient light. Understanding the light produced by decorative fixtures is essential. If chosen more for their art than their light, additional lighting methods are required. A central chandelier, table lamps and wall lights can all provide ambient lighting. Overhead lighting should be controlled separately from the lamps and wall lights to help with layering the lighting effects.

It's important to understand how different shades and light sources will affect the lighting. A bare filament light source, or one concealed in a glass shade, will need dimming to reduce the intensity of the filament. A frosted glass shade or light source will produce a more diffused, softer and flatter light. Bare filament sources create sharp shadows on walls and ceilings (which may be the desired effect). Understanding the exact effect of each light source is vital in the selection of the best decorative light fixtures. It is important they can be dimmed and that the colour temperate is appropriate.

For table lamps and wall lights, the choice of lampshade is important. A diffused or parchment shade will produce the maximum side light. A solid lining will direct light up and down, but not sideways. Sideways light flatters faces, but sometimes the dramatic feel from an up–down light is the desired effect.

Designers often select a combination of shades to achieve different lighting effects. Silver linings are cooler, while gold gives a warmer, reflected light, and a champagne lining sits in between. A critical factor is selecting the LED lamp. A colour temperature (see Chapter 4) of 2,700K is considered warm and may work with a cooler-style interior design or with gold linings in the shades. If a warmer light is preferred, consider using 2,400K, which is similar to a traditional incandescent source.

It is important to consider where lamps are positioned in terms of their lighting control. Ideally, they will be on switch outlets (in the UK, 5-amp outlets). You can then control all the lamps on a single circuit so they all dim together. The alternative, of plugging them into a power socket, will mean having to turn each one on/off individually. I try to add 5-amp or switched outlets to the corners of a room, as it is then easy to add an uplight to avoid these corners being dark. If the seating is central, ensure there are switched floor sockets below the sofas for localised lighting, as otherwise the centre of the room will be flat and dark, as the light is all around the perimeter. If additional lamps are inappropriate, swing-arm reading lights provide soft pools of light on the ends of a sofa; they will highlight fabrics and are practical for reading.

In minimalist interiors, the clutter of decorative lights may be too much, so ambient lighting takes a lead. Downlights are a possible solution but they're usually direct and not so flattering, so may be better used for accent lighting on artwork or a coffee table. Above all, avoid putting in a central grid of downlights: it may light a room generally, but will not create an interesting focus and can be uncomfortable to sit beneath. Better solutions include uplighting, using coves in the ceiling or above joinery units, or wall washing or grazing to highlight texture on a wall. Grazing is not appropriate if art is being displayed as it will light only the top of the painting and send a shadow below.

Figure 9.5

Layers of lighting in this double-height space make it intimate. The art over the fireplace on the left is lit with a picture light, while two downlights focus on the picture under the gallery. At low level, lamps with a diffuse shade give a localised ambient light, while the metal-shade reading light at the end of the sofa lights the sofa fabrics and is a practical reading light. At the upper level, the uplit beams give ambience. Three beams run across the space; the centre one (out of view) has two narrow spotlights that focus on the large coffee table.

ACCENT LIGHTING
IN THE LIVING ROOM

In a living room, there may be accent lighting on artwork and other features to build up layers of lighting and create a balance of lighting. Accent lighting may come from the perimeter of the room – recessed spotlights, track lighting, picture lights or framing projectors.

Consider how to avoid the centre of a room being flat if the lighting is around its perimeter. Putting sockets beneath central seating is a good idea so you can add lamps in the centre of the room. Accent lighting on coffee tables, ottomans or flower arrangements on display tables can also bring light to the centre of larger rooms.

Figure 9.6

This living room combines a desk in the bookcase unit. The soft ambient light is from lamps and from the backlight for the display niche and desk. The general infill light is from the directional wash that lights the bookcase itself.

Figure 9.7

This chandelier directs no light below, so four hidden, independently controlled miniature spotlights are installed within it to focus light on the coffee table. The lamp at the end of the sofa provides task light for reading and accent to the end of the sofa.

A central pendant may not provide enough focus to the centre of a room. I often add two narrow-recessed or surface fixtures with a honeycomb louvre (for extra glare protection) either side of the pendant. A discreet solution, without using ceiling lights, could be miniature spotlights concealed within the pendant. Miniaturisation of LEDs has made this possible. It is best to control these separately with a remote driver.

TASK LIGHTING IN THE LIVING ROOM

CONTROLLING LIGHTING IN THE LIVING ROOM

Good task lighting is often needed for reading in a living room. Directed downlight from the ceiling is not ideal as it is uncomfortable to sit under and can be too intense. However, reflected light from a downlight directed onto a picture can work. Many slim, standard task lights are now available which can introduce another level of light and be effective reading lights.

Successful living room lighting is achieved by controlling each effect separately. This allows the atmosphere of the room to be tailored to different activities and times of day. Chapter 5 has information about how to do this well.

Figure 9.9

Pools of light from the spotlights mounted to the sides of the beams create contrast between what is lit and what is left in shadow.

GARDEN LIGHTING

If there are no curtains and the living room leads out to a garden, then lighting the exterior is a successful way to expand the view. This is effective however small or large the garden.

Figure 9.8

Uplighting the outside wall extends the space. Ambient light is from the linear LEDs concealed in the perimeter slots and the soft shelf lighting gives mid-level ambience. Two downlighters focus on the central coffee table.

CHAPTER 10

DINING ROOMS

Today's dining room may be the heart of a family home, often within the kitchen area, and sometimes part of a living room. As purpose and the mood of the room will vary throughout the day, controlling the atmosphere is essential.

Figure 10.1

The pendant is the visual focus and gives a soft general light, but on its own would be insufficient to light the space. Two spotlights give a soft wash of light on the table, and directional lights around the perimeter highlight the logs and the fireplace and also catch the walls, adding a soft, reflected ambience.

THE DINING
TABLE

The main focus in a dining room should be the centre of the table, which can be achieved in many ways. A pendant will create a visual centrepiece for general light but may not be effective in producing sufficient light on the table. You can achieve a creative focus by putting two recessed spots either side of the pendant. If there is no pendant, a single spotlight could focus on the centre of the table, lighting a centrepiece of flowers or decorative items. I usually use a 10° narrow beam for this, sometimes with a honeycomb glare guard. If recessing in the ceiling is impossible, a surface spotlight may be used.

Sometimes it is possible to integrate a secondary source, like a miniature spotlight, into a pendant, which should be controlled separately from the pendant. This works well if a recessed spotlight is impractical.

Figure 10.2

The central pendant can be articulated to provide downlight or uplight. The key focus comes from two hidden narrow-beam downlights either side. A linear wash by the glazing adds to the general light.

Figure 10.3a
Figure 10.3b

The first image (a) shows a dining room lit with a chandelier and wall lights. The second (b) shows the difference when spotlights are incorporated into the chandelier to highlight the centre of the table. Depth has been added to the shutter box at the window with a narrow 1-watt uplight.

Figure 10.4

Three pendants focus light down and have no need to be supplemented with downlights. Lighting the wine-bottle display adds a mid-level focus.

Some styles of pendant create a downwards focus and no additional light source to the table is required. At night, a stunning effect can sometimes be achieved simply with candles.

PERIMETER LIGHTING IN
THE DINING ROOM

Figure 10.5

*In this relaxed dining area, the key ambience
is from the shelving unit. This uses an up-and-
down technique that allows the objects to
be lit softly upwards as well as downwards.
This technique of lighting joinery when used
as an effect in basements can be like a window
within the room.*

Once the table is lit, consider the perimeter of the room. If there are shelves, their lighting
will provide a focus and mid-level soft ambience. Lighting pictures can have the same
effect. Lighting choices for pictures will depend on the style of the room, the type of art
and how it is framed.

General lighting in the dining room could be from decorative sources such as pendants,
wall lights and lamps. LED lights may be concealed in a ceiling coffer or a recess around the
perimeter of the room or a wall could be uplit from the floor. The colour of the walls is also
a consideration – bare white walls and ceilings reflect a lot of light. So, understanding the
colours of a room and its pictures is important for designing a lighting scheme.

ADDITIONAL TASK LIGHT

Figure 10.6

The general light for this dining table comes from the pendant above. Two independently controlled narrow-beam 10° downlights either side add impact by lighting the flowers.

Sometimes a dining room will double-up as a home office, and extra light may be required on the table. If the answer is simply placing a task light on the table, sockets under the table should be considered at the outset. Another solution is using overhead lights with two different beam widths – narrow for the centre and medium beam either side – to give a more general light on the table. Ideally, these would be controlled separately but even together the intense narrow beam will punch through the medium beam when dimmed.

Figure 10.7

The pendant is the focus – a sculptural piece that acts as an uplight. Directional downlights brighten the table and small eyelid downlights at the front of each shelf provide general light, replacing the need for wall lights. Recessed uplighters to the opening extend the space.

CONTROLLING LIGHTING
IN THE DINING ROOM

Figure 10.8a
Figure 10.8b

These images demonstrate the power of layers of light. In the first (a), the room and table are downlit. By introducing backlight to the marble sculpture in the niche and behind the banquette, a welcome ambience is added (b).

Flexibility to achieve the correct atmosphere is essential in a dining room, as the mood will vary from lunch to dinner, and even more so if candles are used. Controlling the various lighting effects on separate channels ensures that each effect can be balanced to change the mood. Some effects do not have impact during the day but come into their own in the evening.

CHAPTER 11

KITCHENS

A kitchen is often the heart of a family home, where good lighting is essential, not only for tasks like food preparation and cooking, but also to adapt to the kitchen's changing functions throughout the day. Different moods are created with light, to suit busy breakfasts, morning coffee and newspaper breaks, working at the kitchen table and evening entertaining. A galley kitchen needs a different lighting approach to an open-plan kitchen, and the precise solutions will depend on the interior style.

Figure 11.1

Indirect cove lighting to the ceiling provides most of the general light. The tall cabinets at the rear have a recessed slot in front of them so that light grazes down the doors. Decorative pendants float over the island and square double downlights provide task light.

THE GALLEY KITCHEN

Figure 11.2

The uplit central coffer, together with the uplight above the unit on the right, make this galley kitchen seem bigger and brighter. Downlights light the artwork and shine through the glass shelves.

A galley kitchen is narrow by definition, with cabinets on one or both sides. The lighting should create a feeling of space and provide good task light.

Downlighting is a good solution. A central row of downlights may make the space gloomier if they are directed down and not towards cupboards. Two downlights or double fixtures directed to both sides provide a wall wash effect on the cupboards or the wall opposite. This will increase brightness, with light reflecting off the units and the wall.

The next step is to introduce localised task light. This may be lighting under cabinets using individual LEDs or linear strips. It is the essential working light in every kitchen.

For an additional feeling of space and height, and an alternative to downlights, uplight the ceiling from the top of the cabinets to give indirect reflected light.

LARGER
KITCHENS

Figure 11.3

Balloon-style pendants providing focus over the island in this kitchen can be dimmed to reduced glare from the exposed light source and are controlled separately to the square double downlights that provide the task lighting. Additional general light is from the infill directional focus on the cooker hood and shelves. The dark dresser unit is a strong feature, with backlight emphasising its orange interior. Individual under-cupboard lights are recessed into two open shelves.

Some houses may also have a back kitchen for extra cooking, second sinks, dishwashers, etc, leaving the main family kitchen as more of a showpiece and a social area. A kitchen encompasses many moods throughout the day, ranging from bright lighting for breakfast and cooking, to soft and relaxing for entertaining in the evening. The finishes will make all the difference.

Figure 11.4

Task light in this kitchen is from a linear LED at the back of the worktop under the cabinets. Hidden spotlights on a high beam (out of view) are directed to the central island. An additional layer of light comes from the linear lighting below the island, which reflects off the floor.

A large kitchen may have a dining section, a space for sitting and a cooking area. Each could have different elements of general light and task light, so it's possible to change the feel of the room at the touch of a button.

The food preparation and cooking area needs both general light and task light; the dining area will have focused light over the table, potentially a pendant; and a sitting area may need task light and accent light.

The central island is often the kitchen's focus. If the ceiling is high, pendants can lower the apparent height of the ceiling. The style of pendants is a personal choice. If they're clear glass, they should be dimmed as the light source will be visible. A dimmed decorative pendant can be supplemented with downlights, either recessed or surface mounted. In a modern kitchen with a lower ceiling, consider using a pair of downlights or square double fixtures without pendants as a clean, simple and effective solution.

Figure 11.5

The ceiling of the kitchen in this house could not be touched; by slightly enlarging the picture rail detail, an uplight cove was created for the ambient light. Task light is achieved using a marble downstand on the shelves to conceal a linear LED. Over the island, low-glare downlights are incorporated in a custom hood. Uplights are added in the window sills.

TASK LIGHT

Figure 11.6

Task lighting is from individual downlights below the cupboards and shelves, providing a soft scalloped effect. The glazed units with glass shelves are lit with mini downlights. For solid shelves, a vertical LED solution at the front works better. The central shelves have direct light from downlights in the ceiling.

Effective task lighting is essential in every kitchen. Your choices may be dictated by the design of kitchen. The following pages show three methods.

Individual Under-Cabinet Lighting (or Under-Shelf)

Task lighting under cabinets and shelves reduces the shadow created by the general lighting. This can be done either with shallow mini-LED downlights, often with frosted glass, or a linear solution. Shallow downlights are usually recessed into the underside of a cabinet or shelf. They can sometimes give a soft scallop light effect on the splashback behind, which can be an interesting design feature. Conceal the light source either with a downstand on the cupboard or an eyelid shield on the light fixture itself (which is often incorporated), so the light source is not visible. Using individual under-cabinet lights is a good solution with a polished work surface, as the reflection is from a point source rather than a continuous linear strip.

Linear Under-Cabinet or Under-Shelf Lighting

These days, linear LED strips are most commonly used to light under cabinets and shelves. These can be positioned below the cabinet, either recessed or surface-mounted in an extrusion, usually hidden behind a downstand or pelmet detail. Ideally, the LED should have an opal diffuser, to protect the LED and to ensure no reflection of individual dots on any polished work surface below. The location for the linear strip can vary. If the surface is shiny, to minimise reflections, place the light at the back nearest the splashback. Any reflection will be visible on the surface only when one is near the counter. For matt surfaces, these reflections are not a problem and the positioning is less critical, so if a softer effect is required the LED strip can be located towards the front of the shelf.

Figure 11.7

An LED has been recessed near the back of the kitchen shelves, and the result is a sharp visual line. Here, the work surface is honed so there is no reflection of the light source on the surface.

Figure 11.8

This kitchen has a glazed roof and is challenging to light; the task lighting is linear LED under the cabinets and pendants suspended from the roof structure over the island, supplemented by four miniature spotlights.

Decorative Wall Lights and Pendants

Pendants provide a focus over a kitchen island. If they are glass, they are more decorative. However, if a metallic or solid shade conceals the source, and therefore directs the light down, they are an effective task light and will not need to be supplemented with downlights.

Figure 11.9

The wall light here is a design statement and also a task light, projecting over the island. Linear uplight above the cupboard provides ambient light, the linear strip under the cabinet is a task light; and the linear LED under the bar at low level reflects off the bar stools.

Sometimes a quirky wall light can help perform task lighting. A wall light can also add to the effect of under-cabinet or under-shelf lighting. Ideally, all task light should be controlled separately because, when the general lighting is low, or off, task lighting adds atmospheric lighting to the room.

AMBIENT LIGHT

FEATURE LIGHTING

Kitchen ambient or general light follows the principles of the galley kitchen (see the start of this chapter). Integrated downlights or linear LEDs are used to reflect off the cabinets or uplight from the top of the cabinets to reduce the amount of infill downlights required. In a historic or traditional kitchen, decorative lighting is often used to provide ambient light. The design will often dictate the solution.

There are many ways of creating pools of interest in a kitchen and simultaneously adding layers of light. It can be done by lighting within shelving, lighting stools under an island or lighting within glazed cabinets. If the dining area is small and wall lights don't look right, uplights behind a banquette add a pleasing soft layer of light.

Glazed cabinets can be a fun addition and replace the need for wall washing the cabinets. Using uplights in window reveals and lighting artwork also add feature lighting to a kitchen and can be introduced into different scenes.

Figure 11.10

This kitchen uses square double downlights over the island. The dining area beyond has a single pendant and is flanked by two lit shelf units. The perimeter lights are directed to the shelves and a linear LED washes the sink recess with light.

TRANSITIONAL
LIGHTING

Some open-plan kitchens lead directly into the garden. This creates an opportunity for stand-alone external lighting: without curtains or blinds, the glazing becomes a dark mirror at night unless the outside is carefully lit.

Figure 11.11

This kitchen almost extends into the garden; lighting the outside is essential to ensure the glazing does not appear as a mirror at night. Inside, layers of light are clearly seen, and include downlights over the island and multiple feature pendants over the breakfast bar, soft reflected light under the island and an infill perimeter wash of light.

CHAPTER 12

BEDROOMS AND DRESSING ROOMS

The bedroom is our sanctuary of sleep and relaxation. But it's also where we wake and dress. Bedrooms need a soft background light for tranquil late use, get-up-and-go morning lighting, adequate task lighting for reading and for a dressing table, and practical lighting to wardrobes. Lighting needs to adapt to changing seasons. Getting up in mid-summer is different from on a dark winter morning.

Solutions for bedroom lighting are varied and a balance of ambient, task and accent lighting is required for a complete design.

Figure 12.1

This bedroom is striking with its dark, tonal colour scheme. Each lit section makes a difference to the overall balance of lighting. On entering the room, your attention is drawn to the bed. Two downlights light the central artwork and the reflective light catches the pillows, and two directional lights at the end of the bed highlight the yellow throw and reflect off the pale-coloured carpet, which brightens the room. The light from the dressing room at the side expands the feeling of space.

Figure 12.2a
Figure 12.2b

*Discreetly located downlights emphasise the
curtains and the end of the bed (a). Without them,
the centre of the room is dark (b). In both images,
mood lighting is created by the floor lamp and
bedside lamps. The reading lights are positioned at
shoulder height to provide the correct level of light.*

A mixed solution might be as follows. To light clothes well, it's best to use a high-CRI (see Chapter 4) integrated wardrobe light on a door-operated switch, or downlights directed towards the wardrobe. For general or ambient lighting, table lamps and pendants are useful, or concealed linear sources integrated into the joinery, combined with an occasional focused downlight to the end of the bed or reflected off walls, blinds or curtains. Additional solutions include using LED lights in a ceiling coffer, either uplighting the centre or as a perimeter wash of light.

Each room needs to be considered individually for lighting that best complements its style and size.

SMALLER
BEDROOMS

Good lighting can increase the apparent size of a room. If there is no room for a table lamp, a linear LED uplight behind a headboard can provide the softness needed, with a similar effect to uplighting behind a banquette (described in Chapter 11). This and a task light for reading may be all that is required for a practical and attractive design. This may be combined with carefully positioned downlights for the end of the bed or to highlight the opposite wall. Wall lights or joinery lighting are a good solution to create balance in the room.

When considering reading lights by the bedside, a task reading light should be controlled independently from the other lighting in the room. If there is no dedicated reading light, each bedside light should be controlled individually. How the bedroom lighting is controlled, like in all rooms, needs to be planned carefully. General light, whether from a pendant or concealed LED lighting, should be controlled separately from the lower-level lighting of shelving, picture lights and lamps.

The bottom of a table lampshade should be level with the top of one's shoulder when reading in bed, so the light is cast over the book. If lower, one would have to lean out of bed to read. The height of the bedside table will affect the height of the lamp, so plan carefully.

Figure 12.3a
Figure 12.3b

In the first image (a), pendants, task light and reflected light off the curtains provide adequate light, but create little mood. The lighting is instantly lifted in the second image (b) by a soft backlight behind the built-in headboard, which gives a lamp-light ambience, and by two downlights for the end of the bed, to avoid the centre of the room being dark.

LARGER
BEDROOMS

Figure 12.4

Soft lighting contrasts with the concrete ceiling and walls. The decorative pendant and wall light have warm filament lamps which appear less bright because of the uplighting behind the headboard, which softens the effect. A linear LED strip under the window seat gives a soft, low-level wash, adding another layer, and is useful as a night-light when dimmed. Downlights in the drop soffits highlight the window seat and the seat by the bed.

The same balance of ambient, task and accent light is required in larger bedrooms, but with additional lighting to highlight the extra space, perhaps a seating area. Ensuring there is light to the centre of the room is important, and the additional space allows for more mid-level solutions to be incorporated, with lamps or integrated joinery lighting.

CHILDREN'S
BEDROOMS

Figure 12.5

In this child's bedroom, the design is created by decorative items. The lighting design will work well as the child grows older and favourite items change. The central chandelier gives a bright level of light, while discreet perimeter downlights highlight toys. At mid-level, a standing lamp and table lamp provide localised softness.

Children's bedrooms may include an element of fun, such as a starlit sky or colour-changing lights. A child is not young forever, so avoid a solution which may be expensive to redesign when, before you know it, the child has outgrown the design.

Lighting shelves works well: the design solution remains the same yet the objects can change. If bunk beds are part of the theme, try wall lights at each level, lighting shelves at one end or creating detail at the back of the bed to conceal a linear LED strip to light the back wall. In a small room, this helps draw the eye to the back wall of the bunk making the room seem wider and lighting the usually shaded wall of the lower bunk. A fun wall light or pendant is a quicker and easier update later.

Figure 12.6

This bedroom is a child's dream: its height allows for a den on a platform above the bed, where lit shelving and niches either side light the space. Below, the bed is made to feel enclosed and comforting with an indirect linear light to give a wash on the headboard and lighting in the niches either side. General ambience comes from the concealed linear strip in the curtain pelmet and infill directional lights providing a pool of light at the end of the bed.

DRESSING ROOMS

Dressing areas may be part of the bedroom, with wardrobes and shelving, or separated by a divide, or in a separate room. I often integrate lighting into the joinery, and directing downlights towards the shelving, cupboards and drawers also works.

In a small dressing room, the shelving and hanging spaces are often open and, if integrated joinery lighting is too expensive, directional lights may be the best solution.

If there are doors, try adding indirect lighting by lowering the cupboards by 100–150mm and introducing an uplight. Combining an indirect element with the direct lighting helps soften the overall effect. For dressing rooms without doors, the internal lighting becomes the main ambient light, reducing the amount of other lighting required in the room.

When dressing rooms do have doors, the internal lighting should be on a door-contact switch. It is important that the light source is concealed. This can be done with a metal downstand or integrated as part of the shelf construction itself. If the light source is visible, it becomes glaring and more noticeable than the clothes being lit.

Having a different finish at the back of a wardrobe can be a fun element, particularly in a guest room. Lighting can be used to highlight this, for example using an additional LED strip at the back. This works for guest-room wardrobes because they can be relatively empty, and so the lighting is more visible.

If there is no shelf above a hanging rail then integrating the lighting into the rail itself is an option. With adjustable shelves it may be possible to have them slightly reduced by 50mm to allow an LED strip to run vertically up each side, rather than on each shelf. This method makes adjusting shelving simpler.

For dressing tables and full-length mirrors, it is important that the lighting is correct for both the face and body. The most flattering light is not from downlights, which create sharp shadows on the face, but from lights mounted either side. These could be diffuse wall lights, or low suspended pendants or backlight integrated into the mirror itself.

If the dressing room leads from the bedroom to the bathroom, it's useful to introduce a low-level light on a PIR presence detector to act as a night-light.

Figure 12.7

This dressing room, linking to a bathroom, has no doors, so the integrated lighting becomes part of the ambient lighting, reducing the number of downlights required. At low level within the plinth, sensor-triggered floor washers link the bedroom to the bathroom.

Figure 12.8

Ambient light in this dressing room is from a soft uplight concealed by the top of the glazed cabinet. The stool is highlighted by downlights, but the main focus is from the lighting within the wardrobes.

Wardrobes must be well lit. Directional angled downlights can provide reflected light back to the room. If there are no downlights, I would integrate lighting within the joinery. This can easily be done by concealing a 2,700K LED strip in a corner extrusion behind a downstand or integrating it vertically on either side to allow adjustable shelves. The source should always have a diffuser to be homogeneous and ensure the dots are invisible. LEDs can be integrated above and below the cupboards to provide lighting to the room itself.

Integrated lighting in a wardrobe can be useful in eaves as one can conceal the LED behind downstands at an angle. If the storage is open, this becomes the light within the room. An accent detail within the integrated lighting can work well, for example, among a series of display shelves; or a niche can be highlighted to show off a feature bag or pair of shoes.

Another effect is to create a hair-drying and make-up vanity zone within a wardrobe unit, with a backlit or front-lit mirror, using wall lights or pendants either side. Even light from either side ensures the face is modelled without shadow.

CHAPTER 13

BATHROOMS

In some modern houses, bathrooms have transitioned from functional spaces to sanctuaries of relaxation and wellness in which good lighting is essential. The need to be bright and practical by day and spa-like at night can be achieved by using lighting design. Bathrooms respond well to light as they generally use reflective materials. Making the most of the finishes is essential; with good lighting, even simple white tiles can be made to look glamorous.

Figure 13.1

The general light is from out-of-view downlights, two in the shower and one over the vanity unit. The unusual and exotic marble finish is highlighted in the corner niche with multiple mini 1-watt downlights 75mm apart and, around the basin, by a light under the mirrored cupboard.

SAFETY

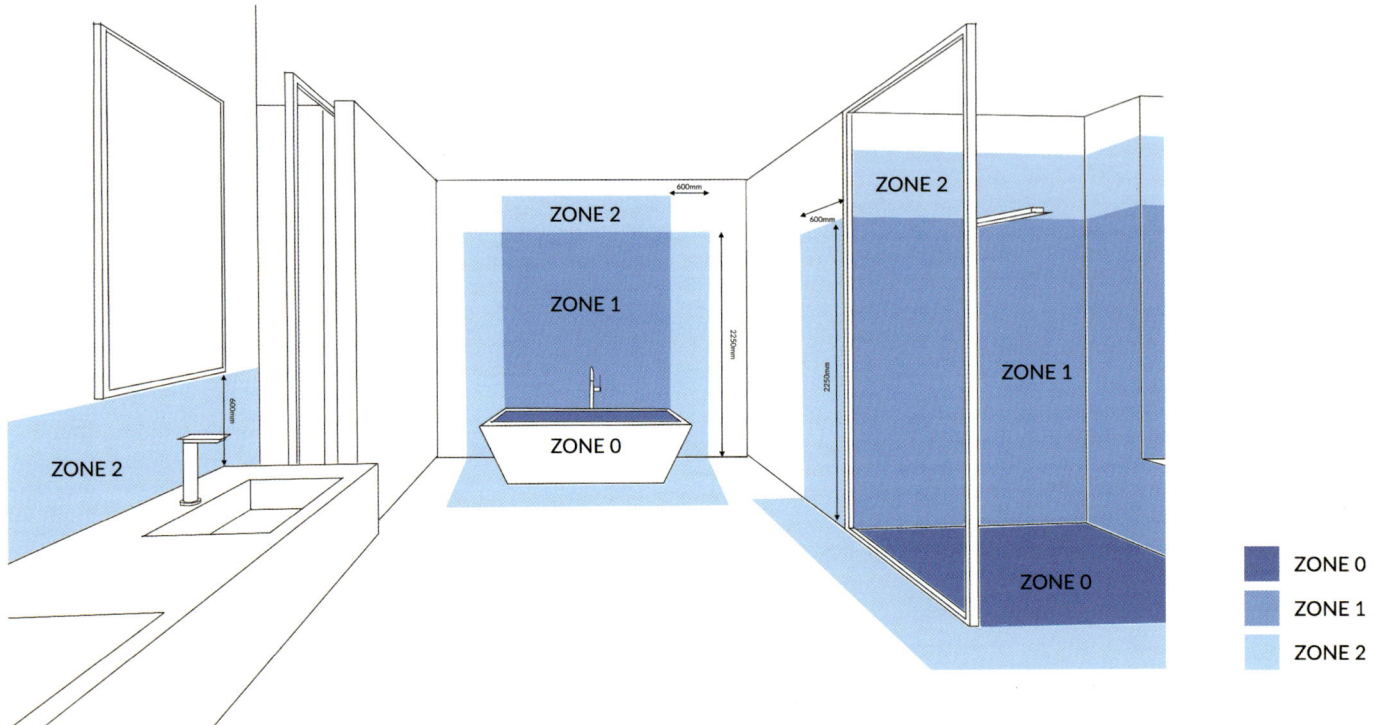

ZONE 2

ZONE 1

ZONE 0

When planning bathroom lighting, safety comes first. Different countries have different regulations. Understanding them is important. This diagram shows the zones around a shower and bath. In zones 0, 1 and 2, lights require a specific IP rating. IP stands for Ingress Protection and there are two figures: for dust and water. In the vicinity of zone 1 (e.g. a shower), a light needs to be IP67 rated and in zone 2 (near the shower or bath), IP44 or greater. IP-rated downlights should be over a shower or bath, and elsewhere are not necessary. This is important because IP downlights tend to have a glass covering to prevent water ingress; this glass covering may pick up dust and condensation, so IP downlights should be limited to where they are specifically required.

Bathrooms must not use mains voltage lighting controls within the room (the exception is in large bathrooms if the bath or shower is 3m from the switch). If lighting controls are wired back to a remote dimmer unit or are wireless, they can be installed within the bathroom. Again, different countries will have low voltage and slightly different codes. In a WC without a shower or bath, the same rules do not apply: mains switches can be located within the room, as can decorative wall lights or pendants without a specific IP rating.

Figure 13.2

Bathroom UK regulation zones.

FEATURE LIGHTING
IN THE BATHROOM

Figure 13.3

*This mosaic wall is lit from above with a concealed
linear LED within a recess. A single downlight lights
the centre of the bath.*

A single narrow-beam downlight over
the centre of a bath is used to reflect a
wonderful pattern of rippling water on the
ceiling and is a popular lighting technique.

Figure 13.4

Low-glare, narrow IP67 downlights are positioned 75mm from the rear marble wall to highlight the bookmatched marble. A narrow slot niche with a mirror back has another downlight to create emphasis, and is controlled separately to allow different mood settings.

An effect to enhance marble and tiles in a shower is to light rear walls using multiple downlights located close to the wall, which adds depth to the space. It's a more interesting option than having a single downlight. The downlights should be close to the wall, and, if possible, set in a recess shadow gap at the back to conceal the light source (a practical solution to conceal an extract grill, too). The downlights provide shafts of light to emphasise any wall texture. You could conceal a linear LED source in a recess, which will provide a soft glow of light. This may require supplementary downlights in the central part of a large shower. If a linear source is used with a grazing optic or a linear grazer (individual units set 50mm apart), a more even downlit wall wash will be achieved. All these approaches help increase the sense of space as the eye is drawn to the rear wall within the bathroom.

Figure 13.5

In this bathroom, perimeter LED lights provide accent and general light, while minimal downlights add infill light and glamour.

Lighting a rear shower wall works well when the tile/stone finish continues up to the ceiling. If it doesn't, consider wall-mounted lights at the top of the tiling or position the downlights out more centrally in the space to avoid the scallop of light appearing above the tile finish.

Using surface-mounted, IP-rated lights in the shower is a good solution if no recess is possible. They are best located just above the shower head to provide a similar effect to the downlight solution described above. This is also useful in a bathroom in the eaves with a sloping ceiling, as the surface lights can follow the sloping line of the soffit. (See Figure 14.4)

Figure 13.6

In this large, high-ceilinged bathroom, the chandelier is out of zones 1 and 2 and provides a general light with an attractive play of light and shadow. A low-level night-light is concealed behind the bath.

The general light in a large bathroom could be from a decorative pendant if it's placed far enough away from zone 2, or it could be from downlights. Ideally, downlights shouldn't be set in a grid but be more precisely positioned over the basin, towel rail and bath, to maximise the impact from finishes and reflected light. Another solution is a concealed cove uplight in the ceiling using a linear LED strip, which gives a soft, shadow-free general light.

MIRROR TASK LIGHTING

Figure 13.7

Wall lights are the best and most flattering way to model the face. The basin and towels are lit by concealed linear LED strips, ideal to use as night-lights when dimmed.

The most important task area in a bathroom is probably the basin. Strong light is needed for shaving and applying make-up. The best way to achieve this is wall lights with frosted glass or a diffuse shade either side of the face. This gives an even, flattering light. Never place a single downlight above the basin. It will create shadow on the face but, if used in conjunction with wall lights, it can be an effective approach. Two downlights either side is better than one in small bathrooms, to add glamour to the surface, where wall lights cannot fit. They are more effective if the vanity unit is white as the reflected uplight will fill in the shadows on the face.

Figure 13.8

Task light to the face comes from pendant wall lights either side of this large mirror. General soft light comes from the indirect up-and-down LED behind the mirror and a further layer of light from an LED is added into the shelves below. Shower downlights positioned close to the rear wall highlight the tiles and add depth.

Another technique is to use backlights behind a mirror. This gives a halo of light around the perimeter of a mirror, which highlights any detail. The viewing angle is important as the LED strip should be hidden. It is best used when contained either side by a wall and light is reflected back towards the face.

Polished tiles in bathrooms may reflect the light source, which otherwise would be concealed. This looks cheap and can be easily avoided. Choosing LEDs with opal diffusers and carefully positioning them to avoid direct reflections is best. For example, with a polished floor and LEDs integrated under a bath or basin unit, rather than point the light directly down, reflect it off the back wall. Also, ensure everything under the vanity unit is a dark colour to avoid reflections of the lights there, if placing lighting beneath it.

ADDITIONAL
ACCENT LIGHT

Figure 13.9

A linear LED is concealed in a recess at ceiling height across the shower and bath. The downlights for the shower are located in the same slot, and one is centred over the bath. The niche in the shower has a small downlight in it – the best solution for small niches – while the long linear slot works best with a linear LED source.

An additional layer of light can be used to help create a spa-like mood and a touch of magic. This is usually achieved with low-level accent lighting that's deliberately not functional – lighting, for example, in a shower niche, an uplight to the bath or introducing lighting to glass shelves. Low-level floor washers are also ideal for both mood-setting and as a night-light on a low setting linked to a PIR, coming on when someone enters the room.

Another way to add layers is to use uplights, potentially to a feature wall behind a bath or to uplight shutters and windows. Miniature low-glare 10° uplights suggest candlelight at night, adding to the spa-like feel.

GUEST
WC

Figure 13.10

The backlight from the mirror shows off these colourful walls. When lighting for make-up is required, the backlight frosting to the front is an excellent facial light. A downlight adds sparkle to the vanity area, and low-level LED strips give soft reflected light and ensure no dark shadows.

Some clients like to experiment with design in the guest bathroom. Good face lighting as well as general lighting are still needed.

CHAPTER 14

UNUSUAL SPACES

Many houses have some awkward spaces – areas under the stairs, a restricted attic or roof space, or a dark basement. All may be useful but can be tricky to light: they may offer no flat ceilings or not enough flat wall space to put joinery or furniture against or to fix a wall light on.

Figure 14.1

In this unusual duplex at the top of a tall townhouse, maximum height has been achieved by exposing the rafters, and a mezzanine level added. The backlit shelving units are a visual link between levels, and intimacy is created in the seating area with table lamps and a focus on the coffee table, which comes from narrow-beam spotlights mounted on the side.

ATTICS

Attics often have low or awkwardly shaped ceilings. Sometimes, they are combined with the floor below to make magnificent spaces where the roof structure can be celebrated rather than concealed with a flat ceiling. Exposed beams provide their own lighting challenges. They can be lit with a linear LED on the horizontal beam to provide a soft uplight with a diffuser. Ideally these should be routed in so they are flush with the beam. If this is impossible and the extrusion for the LED is shallow enough, they may fit on top of the beam and still be out of view. Spotlights could be mounted on either side of the beam, uplighting the angled woodwork above. They can also sit discreetly on a horizontal beam and focus light down, possibly on pictures or a central coffee table. If lighting down from the beam, always ensure the light fixture is on the side and cannot be seen below the line of the beam, so the clean architectural visual lines remain intact.

Figure 14.2

Spotlights mounted on each end of the beam emphasise the architectural pitch of the roof truss. Spotlights downlight the central coffee table; these are mounted to the sides of the beam and are never visible below the bottom of the beam, so do not distract from the cleanness of the structure. Lamps and joinery lighting provide infill and accent lighting.

Figure 14.3

This freestanding bath, set in the eaves with a dormer window above, has two floor washers behind the bath at each side, which reflect light off the bath to give a soft glow that is ideal as a night-light. The dormer window has 1-watt narrow uplights that frame the shape, and also a simple downlight in the only flat soffit, which lights the centre of the bath.

Figure 14.4

Here, a sloping ceiling makes directing downlights difficult. The most practical solution is to use IP-rated (see Chapter 13) wall-mounted fixtures as close to the ceiling as possible, with the light glancing down the tiles.

Beam structures may not be visible in some attics, only the sloping soffits, ideal spaces for bedrooms or bathrooms.

An attic bedroom with a sloped ceiling can be well lit with table lamps. Low-level, plug-in lamps or sculptural floor lamps are perfect for low, sloping eaves. I usually try to include a downlight into a dormer window which in daytime brings in light, but at night falls into shadow. By lighting the blind, the window is brought back into the room.

Bathrooms in the eaves also have their challenges. There are plenty of solutions, from low-level uplights to surface fixtures in showers that can create interest.

Figure 14.5

An attic can make a good study or office. This one benefits from the end window, providing plentiful natural light during the day. General light is created by integrating three lines of linear LEDs into the top of the shelving; covering them with a frosted-glass top creates a diffused ambient uplight in the room. At night, two desk lamps provide task light.

UNDER THE STAIRS

Figure 14.6

A linear strip is recessed into the floor for a soft upward light and combines with the lamp to bring focus to this under-stairs area.

A storage cupboard is a common solution for the space under the stairs. It can be opened up to become a work space or a display area and, instead of being blocked off, can be transformed by clever lighting into an area of interest.

The under-the-stairs space can be brought into a hallway, to widen the feeling of space, by uplighting from below and displaying a sculpture or table with a lamp on it. By lighting these forgotten areas, they are brought into a room, creating space, openness and width rather than being blocked off and only used for storage.

Figure 14.7

This kitchen tucked away under the stairs, was dark, so by adding lighting to each shelf from the front providing good task light and visual interest while pendants focus on the island.

VAULTED AREAS

Figure 14.8

The halo-lit mirror in this basement bathroom provides the main lighting, which reflects off each side. This indirect light helps light the face, while accentuating the barrel of the ceiling. The small square floor washer adds emphasis to the entrance and the LED strip below the vanity unit continues the soft ambient light.

Vaulted basements can be used as downstairs WCs or, if large enough, for a wine cellar, a dining area or even a kitchen. The size of the arch and how much general light is required will dictate the lighting solution. Traditional approaches will usually be difficult: a brick arch cannot be perforated after tanking and waterproofing. Basement electrics must be considered early. Let's explore some of the solutions.

Figure 14.9

This low-barrelled ceiling where no lighting fixtures could be recessed feels glamorous with two pendants. If these were the only source of light, one would have been drawn to the visible lamp in the pendant, but this is diffused and softened by the backlight behind the mirror to provide a soft halo. An uplight at floor level highlights the curve of the arch. Three different lighting effects have created a glamorous space that could otherwise have been forgotten.

Figure 14.10

To have a good level of light in this kitchen, and make the most of the barrel arches, the ceiling lights needed careful consideration. A tray system at the base of each barrel arch houses an LED uplight to the vault, providing general background light, together with the linear glass pendant above the kitchen island. Spotlights mounted below the metal tray system on the ceiling provide task light. Additional light layers above and below the kitchen counters and under the island create the balance.

WINE CELLAR

Figure 14.11

In this wine cellar, two downlights focus on the tasting area, the wine racks are backlit, adding depth, while miniature downlights catch the front.

Basements are often used for wine cellars. There are many lighting effects to make them fun and interesting. Bottles can be lit with linear light or miniature spots to highlight their labels. Backlighting can be successful, either on each shelf or, if the wine rack is pulled away from the wall, a grazing technique to light up the rear wall.

Wine cellars need to be temperature controlled. If wine coolers are required, the lighting needs to be discussed with the manufacturer. If bought off-the-shelf, the lighting can be a very cool 3,500K, which will look incompatible with the rest of the space. Bespoke units with the lighting at a colour temperature of 2,700K, to match the rest of the LEDs in the room, is ideal.

Figure 14.12

This vaulted basement has been transformed into a wine cellar and a magical dining room. A soft ambience comes from the chandelier but this is dimmed and the drama is created by the narrow uplights that highlight the brick walls and arched brick ceiling.

CHAPTER 15

LEISURE ROOMS AND POOLS

Basements sometimes accommodate cinemas, pools, gyms, wine cellars, bars or games rooms. They are, of course, a luxury and can be further enhanced with creative lighting. Lighting can do more to reveal or conceal and change than any other design element. Depending on budget, if the finishes are simple, light can transform any space at the touch of a button.

Figure 15.1

The success of this minimalist pool room is its relationship with the outside. The glazing opens to a garden and the external lighting becomes the most important element. The pool itself is lit on one side only, from the far wall, to avoid seeing the light source. The columns are illuminated from downlights positioned very close to them, and there is some infill light from downlights between them, on a separate circuit.

SWIMMING POOLS

When lighting an indoor swimming pool, it is best to minimise the number of downlights, particularly over the water, and to keep the lighting to the perimeters. This is for several reasons. First, it's to limit difficulties of access where there is the need for any maintenance over the pool itself; second, it allows for a clear, uninterrupted ceiling space; and third, when doing backstroke, you don't want to be blinded by downlights. Water reflects light and can be your main ambient light. The most successful pool schemes use downlights sparingly and combine them with other effects, such as lighting beneath the water, illuminating a textured wall and accenting other structural elements or materials. Each area of the pool may require a different solution. Seating and any artworks should be highlighted to bring together the different elements of the design scheme.

Figure 15.2

This lower-ground-floor pool is successful partly because of the view to the sunken courtyard, with its green wall. It illustrates the value of lighting outside areas to create a feeling of space and depth. The ceiling is clear of downlights. Instead, there is lighting withing the pool itself, and columns are uplit. The end wall is strongly uplit and downlit from a recessed perimeter slot.

Figure 15.3

This basement pool has no natural light, so the effects are totally artificial. The concept is about playing with light on layers of contrasting materials. The timber wraps across the ceiling and floats off the wall, accentuated by linear 2,700K LED lighting between the layers. At each end, custom light-art pieces are introduced to create the focus, reflect off the water and contrast with the simple palette of materials. Even the bench is a piece of sculpture in the space.

How one plans to light a pool area will change depending on how the pool is to be used and whether there is natural light, as this influences moods throughout the day. A children's play area during the day and a spa pool at night need to be treated differently.

Figure 15.4

The challenge in this pool is the striking folded ceiling and the desire to keep it free of lights. The perimeter lighting had to be bright enough to be practical. The play of light on the ceiling comes from reflected light off the wall, which softly emphasises the folded panels using light and shade: some are lit and others remain in shadow. Colour has been introduced to make the space fun, to contrast with the neutral, polished-plaster walls. At the far end, a skylight floods light in during the day, and at night, a linear flood using a cool colour temperature of 3,000K emulates that effect. The pool floor can be raised to provide a shallow kids' pool or even a dry flat surface for a party, so the lighting selected within the pool is multiple small sources rather than a few larger pool lights. The small fixtures are at a higher-than-normal level, so they are effective when the pool floor is raised. (See figure 4.1).

Figure 15.5

This swimming pool is glamorous and subtle. It plays with various elements of light. Minimal downlights highlight the end of the loungers; a continuous uplight to the perimeter of the pool uplights the walls and this effect continues behind the loungers. The lighting changes with the panelled walls, divided by bronze rods supporting robe hooks, which are individually downlit with a defined scallop of light. Finally, the drop pendants (custom-made to be IP-rated) between the loungers bring intimacy and introduce another layer of light to help build the overall mood. Each effect can be controlled individually and preset for the best balance.

When designing a pool area, always check lighting codes to ensure that equipment selected is suitable for water. IP ratings (see Chapter 13) will show how water resistant a luminaire is. Chemicals used within the pool may be corrosive and must be checked against the choice of fixture. Sometimes a marine-grade stainless steel is best. Ozone and other softer cleaning agents make pools a less corrosive environment and so you widen your choice of fixture.

CINEMA ROOMS, BARS AND OTHER AREAS

Whether a cinema room is a wall-mounted TV or a dedicated large screen with surround sound, the lighting should be designed for different moods. Perhaps design a bright scheme with background light to make an atmosphere that's perfect for sports viewing. But also create a low-light setting for watching a romantic film. These rooms can be dark and dramatic. Acoustics, comfortable chairs and mood lighting are key.

Figure 15.6

This cinema room uses a flexible LED strip for an indirect soft light in the circular ceiling details, and also in the long orange recesses behind the sofa, to give a local level of light. The extra integrated light in the base of the coffee table is ideal for the lowest-light-level scene.

Figure 15.7

In this basement bar, a mirror ball reflects the room and, with a spotlight at night, creates a party atmosphere. The rear bar, with its integrated shelf lighting, is the main feature.

Figure 15.8

This is the ultimate large cinema, with circular pendants in the coffers, wall lights and individual lamps by each seat for a sense of intimacy. Small, recessed spots send light across each step.

BOWLING
ALLEYS

An ultimate luxury is putting a bowling alley in a basement. The lanes are lit to provide focus along the track. Coloured neon art and ultraviolet light add further interest.

Figure 15.9

The entrance has a fine collection of backlit balls and boots, with directional downlights lighting the front. A lamp gives a local layer of light by the seating.

Figure 15.10

The drama of the alley is created with narrow-beam lights to the lanes and multiple candle-like wall lights creating an impact on the back wall, which contrast with the ultraviolet light on the bowling pins.

PART 3

CASE STUDIES

CHAPTER 16

CONTEMPORARY TERRACED HOUSE

THE LIGHTING BRIEF

We were commissioned to devise a complete lighting scheme for this terraced house as it was being totally redesigned and refurbished to create a contemporary home. All the original internal walls were removed, and a double-height, highly glazed extension was added at the rear, housing a kitchen and dining area. The lighting brief was to reflect the new contemporary style. Challenges included bringing life to long narrow spaces and double-height areas, as well as harmoniously linking the old and new sections of the house. The owner's artworks needed expert lighting, to bring them to the fore of the interior design scheme.

Figure 16.1

Layers of light draw the eye around this open-plan, double-height space and through the lower opening.

THE SOLUTION

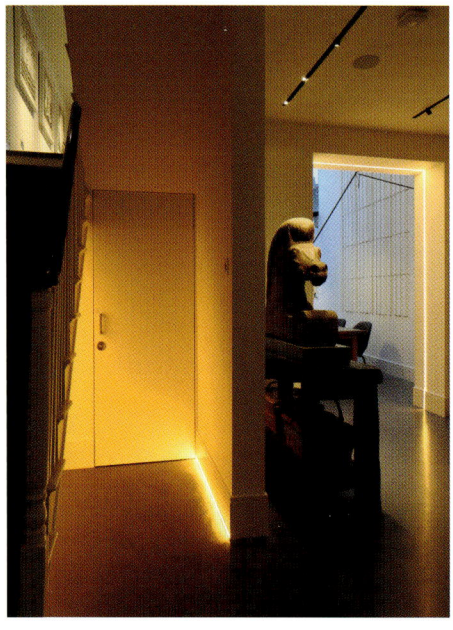

Figure 16.2

Layers of ambient light are provided by the linear strips at floor level defining the opening to the kitchen. Accent light is focused on the sculpture.

Figure 16.3

The horse sculpture is prominently lit with two 12° spots to ensure the head is modelled from two sides.

On entering the house, we are greeted by the end of the original wall separating the staircase from the main living room. Removing this corridor immediately makes the terraced house feel wider. The awkward under-stair area is lit with a dimmed LED strip concealed behind the stair string, making it part of the room rather than a forgotten zone. To contrast the soft under-stair lighting, another LED strip is recessed into the base of the skirting board, providing a graphic linear wash across the floor, leading to the door of the downstairs cloakroom. Lighting is a useful device to increase the feeling of space and create focus.

In the main room, a recessed magnetic track housing spotlights gives flexibility to the lighting scheme. This is a useful way to light artworks, particularly if they may change or move. The lights can be easily adjusted to sit flush within the track as a downlight or pulled out as an accent light to focus on the stone horse sculpture. Beyond is the entrance to the dining area, off the kitchen, and this opening is delineated with another graphic LED line of light.

Lighting the horse-head sculpture with two spotlights from the track above particularly highlights the texture of the mane, giving life to the intricate stonework. The light also creates a play of shadows, exaggerating features such as the mouth and veins, adding strength to the sculpture.

The upper-floor rooms offer a view into the double-height space through linear-LED-lined openings, which were once windows. Light draws your eye up to these rooms from below, and when you're in them, out into the open-plan space. Soft lighting is created with simple angled downlighting in the rooms, as well as task light in the study and the use of a multi-angled wall light in the living room.

Figure 16.4

The double-height space, with a fully glazed roof, means there is no opportunity for mounting lights, creating a challenge for the lighting designer. The wall-mounted light is a solution, offering essential task light to the table. The openings are defined with a warm 2,400K LED line for soft ambience and to reinforce the dramatic height of the space.

Figure 16.5

In the kitchen, a linear LED strip is concealed at the back to softly wash light down the cupboards for general light. Over the island, recessed square double lights give task light to the work surface and shelves.

Coming into the double-height volume, we can appreciate how the glazed extension has filled the previous void between the house and the neighbouring wall. By defining the new openings with LED lines, we accentuate them and draw our focus upwards. These layers of light offer essential evening illumination.

The texture on the wall looks like a modern art installation but is actually a set of panels added to solve the acoustics of the space. The angled wall light, needed as no lighting could be attached to the glazed ceiling (out of sight in the image), can be adjusted to focus over the table, ideal for reading a newspaper, working at a laptop or having supper.

The kitchen area has a low ceiling so the line of cupboards, fridges and ovens are washed from the ceiling with a linear LED strip. This has been concealed within a slot to provide a soft ambient light, which also reflects general light to the room.

The island unit needs strong task lighting. For this, we used recessed double downlights in white with a medium distribution, with one set directed towards the shelves. The background light is soft from the concealed linear strip, and downlights are required only for task light over the island.

Figure 16.6a
Figure 16.6b

The kitchen area looks out onto the garden. When the garden is not lit, the window is a black hole (a); when it is lit (b), the eye is drawn out, connecting inside with the outside and creating depth.

Figure 16.7

The sculpture is lit from above with a directional spotlight.

The problem with a totally glazed box is that the reflections at night are of the inside; the glass becomes a black mirror. The solution is to light something beyond – in this case the small garden. The dramatic red wall forms the backdrop to a bronze sculpture, which we lit from a high-level spotlight and put into silhouette with recessed uplights set into the decking.

The drama of the uplights on the red wall and the uplights on the textured pots contrasts with the spot-downlighting on the sculpture. Likewise, the LED strip under the top of the raised planter washes the ground, while the tree is uplit. Just as with interior lighting, it is the layering of effects that makes an exterior scheme stronger.

Figure 16.8

Multi-pendants in this narrow stairwell send the light in varying directions, so play with the space.

At the top of the stairs to the bedroom there is a contemporary, multiway pendant that fills the void. Uplights light the tall dormer window that leads to the roof terrace beyond. At night, leaving just the uplights on gives enough light to navigate the stairs.

In the bedroom, a skylight above the bed allows the sky and stars to be seen and enjoyed. It is defined by an LED strip and this graphic line delineates the skylight. It has an opal diffuser, so the direct reflection in the skylight works as an effect and, when the blind is closed, the skylight will act like a coffer, reflecting the LED line off the blind. The wardrobe opposite uses linear LEDs below and above to add general light to the room. At night, the lower linear LED can be dimmed to a very low level to act as a nightlight. The gap to the ceiling or floor needs to be a minimum of 100mm to provide a soft wash effect rather than too harsh a line.

Figure 16.9a
Figure 16.9b

In the bedroom, bedside wall lights are perfect for reading. The ambience was more challenging. By day, natural light floods in through the skylight; at night, we made the recess glow with a diffused linear LED.

THE VERDICT

The lighting has been concealed to provide lit spaces that are contemporary, simple and uncluttered. Integrating details and minimising downlights to task and accent only is a strong element of the design. The balance between inside and outside lighting is also important, allowing the garden to flow in and give an added perception of depth from the house.

CHAPTER 17

TRADITIONAL TERRACED HOUSE

THE LIGHTING BRIEF

This project is in a listed building, with all of the strict controls that this entails. Our aim was to respect and highlight the original fabric of the building and to work very closely with the interior designer to establish the right approach that complemented chosen palettes and finishes. We deliberately minimised the use of recessed downlights, except on the upper floors. The overall feel is classic/contemporary, while accentuating original architectural features with light. Many of the timber and stone floors are new, allowing us to incorporate uplights. The lighting concept is achieved by layering decorative sources, extensive joinery lighting (particularly in shelving) and infill architectural effects, including surface-mounted spotlights and recessed uplights to door and window reveals, fireplaces and panelling.

Figure 17.1

Balance is created between decorative and architectural lighting effects.

THE SOLUTION

Figure 17.2

This backlit shelving, framing the entrance to the kitchen from the breakfast room, brings light to this side of the room and draws you through the doorway.

In the entrance hall, the main ambience is from the central pendant and decorative wall lights. Additional emphasis is given to the architecture using low- energy uplights to door, architrave and window reveals. This contemporary addition adds general light and brings out the period details. Out of view, two surface spotlights either side of the pendant focus on the flowers on the hall table. Another solution, depending on the decorative light selection, is to conceal small spotlights within the lantern or chandelier fixture itself to light the table below.

The entrance from the breakfast room to the kitchen is framed by shelving units, glazed on two sides. Our successful lighting solution conceals a vertical LED to the back wall of the shelving, creating two columns of light, putting the display into silhouette. It helps provide a soft balance of light to the room, which has pendants over a table and two wall lights. Without integrating lights in the units, this area of the room would be in shadow.

Figure 17.3

View to entrance hall from living room and the table lit by two spotlights either side of the lantern.

Figure 17.4

In the evening, the mood is altered with warmer uplighting, and other light sources are dimmed.

Lighting in the kitchen was challenging as we were unable to use the ceiling except for decorative lighting, because it was listed. The design solution, developed with the kitchen designer, was to extend the shelf above the units to incorporate an uplighter, giving the ceiling a soft wash of light. We used a linear LED with a cooler colour temperature of 2,700K during the day and a warmer colour temperature of 2,400K for a softer evening effect. This was achieved using a tuneable LED strip, but can also be done with two LED strips on separate channels that can be controlled individually to get the perfect colour balance. In addition, the thick shelf houses small 4-watt LED downlights to light the front face of each cupboard door, increasing the light on the surface below. Under the cabinets, a linear solution provides task light.

Creating an upward light from cabinets is a good way to create a soft, shadow-free general light for the whole space. The task lighting can then focus on the counters. This solution also reflects light off the ceiling so is ideal in rooms with high ceilings, as fewer downlights are needed.

Over the bar counter, three pendants with a soft linen shade create focus. The shade is open at the base, allowing a softer downward light onto the counter. An LED strip recessed below the counter top lights the stools, adding to the layers of light in the room with a soft reflected light from the stools and the floor.

By the window, two wall lights provide both ambient light and task light to the sink, with two recessed uplights on the shutters offering accent light.

Figure 17.5

Tailored front-lit bookcases frame the entrance to the library. It's a shelving light solution that is often used in more traditional-style houses. Backlit shelving doesn't work with books as the titles need to be read that is often used in more traditional-style houses.

The study and library walls and timber shelving are dark. The shelving frames the doorway between the spaces and is lit with a concealed linear front light. This softens the look, and is an ideal way of adding gentle mid-level lighting, as well as giving accent light to both books and objects. The white fireplace, emphasised by 1-watt uplights set into the hearth, is the room's focus, the lights drawing the eye even when the fire is not lit.

Figure 17.6

In this bright living room, the white interior of the shelving helps to reflect the light. Uplights in the shutter boxes add depth, and the uplit fireplace gives focus to the room. The chandelier, mid-level wall lights and picture lights complete the layers.

The living room has a much lighter colour palette and feels bright. The focus is the crystal chandelier, dimmed even on the bright setting to avoid the lamps creating glare. The shelving's pale background reflects far more light than the darker shelves in the library. Considering the colour of shelving is imperative, as it changes how light works. Accent lighting comes from picture lights to highlight the art, and uplights for the fireplace and shutters. Lamps and wall lights add visual focus and mid-level light. By controlling each element separately, different moods can be achieved.

The small bedroom has a bedside light for ambience, in addition to four downlights, two lighting the headboard and two throwing light to the base of the bed, a useful technique to bring light into the centre of the room when ceilings are too low for pendants. Downlights can be more discreet in low ceilings, particularly if deeply baffled, so that the viewing angles are shallower and you're not staring at the light source. The downlights to the headboard offer a focus when you enter the room, and an appealingly glamourous look, but they should be controlled separately so they can be switched off independently when in bed, when reading lights are all that is needed.

Another trick that's useful in attic bedrooms is to add a light in the dormer window, which highlights the blind and increases the feeling of space, as this area is usually in shadow at night.

In the centre of the cupboards opposite the bed is a small desk area. This allows lighting to be integrated into the joinery, both to the shelves and with a strip of LED under the desk, to catch the chair and create an inviting feel to this bedroom and ensure this end of the room is not dark. Practical lighting inside the cupboards is from a linear LED, concealed by a downstand to avoid direct view and operated on a door contact switch.

For the passage from the master bedroom to the dressing room and bathroom, uplights at the doorways to the dressing area and bathroom add a sense of drama, and mean fewer downlights are needed. They're also useful as nightlights when dimmed.

Figure 17.8

This elegant bathroom is lit more like a reception room, with a chandelier for ambience. The shower is given extra attention with concealed downlighters, which shaft light down the marble wall.

In the bathroom, the chandelier provides glamour, as does highlighting the marble in the shower with concealed downlights set within a ceiling recess. This keeps the fixtures out of view and is also useful to hide the extract grill. Wall lights either side of the vanity unit are the perfect task-lighting solution for the face. A small night light below the basin, controlled automatically on a presence detector, is ideal.

Figure 17.9

In the dressing room, the pendant draws your eye. But much practical light comes from a subtle lighting of the front of the cupboards from recessed downlights, which widen the room and also softly highlight the cupboards' design.

The beautiful dressing room cupboards are softly washed with scalloped light. Softening lenses are applied to the downlights to smooth off the edges of the beam. This adds to the practicality of seeing the contents of the cupboard when the doors are open. Reflected light from the doors is the main source illuminating the room.

Figure 17.10

For this narrow passage, more than just downlights was needed. Uplights help create drama, and increase the light level.

The decorative lantern helps with infill light, as well as being a focus to distract from the downlights directed onto the cupboards. Interest is given to the room by creating a sculpture niche, which also breaks the long run of cupboards. This has two recessed uplights and a directional spot onto the sculpture.

THE VERDICT

This house has a calm ambience, created with a flowing palette of muted colours and the carefully planned lighting scheme. This project had many challenges owing to the listed status of the building, meaning that many ceilings and architectural features could not house lighting. The successful solution was a harmonious combination of decorative lighting in keeping with the traditional sensibility of the interior with much concealed, built-in joinery lighting and strategically located architectural uplights in the flooring.

CHAPTER 18

BARN CONVERSION

THE LIGHTING BRIEF

This spectacular and unique home is a tucked-away secret. The approach is through a small, densely planted front garden concealed from the road by a high wall. Its simple two-storey, barn-like structure belies more exciting features, including a riad-like internal courtyard – a jewel to be discovered. The whole residence has a tranquil sense of calm which is reflected in the lighting design. The building's pitched barn roof runs the full length, and the deep plan means that the central courtyard brings in vital daylight. Glazing on both levels brings wonderful changing views of the courtyard. Lighting it from all angles was a wonderful challenge. The house also features dramatic use of contrasting colour, which required special attention for lighting.

Figure 18.1

An open central courtyard is what this deep-plan property pivots around, bringing natural light to its heart by day and a theatrically lit magic by night.

THE SOLUTION

Figure 18.2

The entrance hall and top of the stairs is lit in layers, with a pendant, a spotlight in the skylight and a lit niche in the wall, plus a downlight in the window recess.

Figure 18.3

For this simple staircase to the lower ground floor, light pools from the spotlight in the skylight and from reflected light in the niche at the top. Despite the dark timber, the contrasting bright cabochon floor reflects plenty of light, which lifts the space.

The entrance, with its pitched ceiling, has a decorative lantern as the general light and a picture light creates a focus on a painting set into a niche above the stairs. Reflected light from this is enough to light the stairs below. To balance this on the opposite side is a lamp on the side table. The two slot windows, each with miniature downlights, frame the front door at night from both outside and inside. Each effect is controlled individually and each adds a layer of light, so the mood can be set to transform the space throughout the day and night.

Figure 18.4

Uplighting the pitched roof structure in the open-plan living room offers general light, and highlights the architecture. The beams house directional spotlights on each side, tilted towards shelving, art and the ottoman, bringing brightness to the centre of the room.

From the hall, one enters the open-plan living room. This has a pitched ceiling, so for general light, a linear LED was concealed on top of each beam to give a soft wash of light to the ceiling. The colour temperature of 2,400K was selected, as its primary use is at night and a warmer, softer light is preferable as an infill light to the lamps, which provide the visual context.

On the sides of the beams, jack-point sockets disguised as bolts are installed, allowing spotlights to be plugged in where required (if not used, they just look like bolts). This provides a very flexible system on both sides of the beams. The spotlights are directed onto the paintings and sculpture, using different beam widths: medium for art and narrow for sculpture. Two are located near the centre of the beam opposite the fireplace and angled to give a glow on the ottoman, creating a central focus to the room. These are controlled separately to the perimeter spots so a stronger pool of light can be created in the centre of the room, and to allow for different mood settings to be chosen.

Figure 18.5

Lamps appear to do all the work on the sideboard, but spotlights help out.

Figure 18.6

Certain niches in the shelving are front-lit, to show off objects such as pottery.

On the sideboard it almost appears that the lamps are doing the work, even though it is the small, narrow spots on the side of the beam that add the contrast. Shelving either side of the fireplace is primarily for books, with specific niches created for special items. To keep things simple and not over lit, the spotlights on the sides of the beams wash light onto the book shelves and the objects are lit with a shallow-eyelid downlight integrated into the selected niche. The eyelid is a drop-down shield used to conceal the light source, chosen because most very shallow downlights can be glaring unless specially designed. This downlight must be near the front of the shelf rather than centred in the niche to ensure a wash of light from the front. The often-misused central position works for a glass object, but is useless for solid objects as they won't be lit at the front. This solution of lighting random niches gives the shelves character and a more relaxed feel than if all the shelves were lit.

The narrow-beam spotlight provides a play of light and shadow on the sculpture and picture, bringing them to life, and the infill lamp light makes the effect appear more natural.

The key to all of these effects is to introduce light at different levels, to achieve layers that are controlled individually to create different moods.

Figure 18.7

The link between living and kitchen spaces looks over the central courtyard. The purpose-built study area provides most of the lighting. At night there is added interest from the lit courtyard.

Figure 18.8

The courtyard, a showpiece at the centre of the house, needed to shine from every angle. We lit the leaves and trunk on the giant fern, and highlighted the sculpture's ultra-smooth surface. The floor is also an artwork, and its 3D qualities are brought out with spots from above.

The corridor study area links the living room to the kitchen, with the rich burgundy wall a dramatic backdrop. To light the area and keep the ceiling free of downlights, the shelves are lit with the same shallow-eyelid fixture used in the book shelving. Unusually, the top shelf uses the same fixture as an uplight to ensure the objects on the top shelf are lit. These effects provide task light on the desk, as well as lamp light for general ambience. This area looks onto the central courtyard, with its ferns and sculptural fountain rising from a custom-laid marble floor, both designed by Jordi Raga Frances.

The central courtyard is the jewel in the centre of this home. Plants are uplit with a combination of narrow spots to the ferns (to project shadows on the rear wall that are visible from both levels) and softer spiked floods for the other greenery. The magnificent granite sculpture is lit with uplights around its base, set among small marble stones. This allows for final adjustment so the uplights can be exactly located to catch the running water. At roof level, two narrow spots shine down to catch the detail of the floor and also the top of the sculptural water feature. Looking down onto the courtyard, the intricacy of the downlit undulating marble floor can be appreciated. Just as in the house, there are layers of lighting to add different impressions depending on where the courtyard is viewed from. The varying heights and textures of the striking features in the courtyard help make it so successful.

Figure 18.9

The pitched ceiling continues in the open-plan kitchen/dining area, and is lit in the same way as the living room, providing continuity. Spots on the shelving offer mid-level light and task light. The raised section in the island provides separates the dining area, and is a useful source of task and general light – a linear strip is concealed at the top.

The kitchen/dining area leads onto a small garden beyond. The pitched ceiling uses the same solution as the living room, with the uplight approximately 300mm short of the ends of the beam, to ensure the flare of uplight is not too intense. The lighting effect reinforces the pitch of the ceiling and creates ambient light.

Figure 18.10

Unlit by day, at night the shelves in the kitchen come alive.

Task and feature lighting either side of the hob is incorporated into the shelves with the small shallow-eyelid fixture and the top shelf is backlit using a linear LED that also catches the sloping ceiling.

The extended bar separating the dining area from the kitchen incorporates a linear strip under the higher timber worktop, providing task light to the counter. A lamp gives a soft ambience between the kitchen and the dining area. The central island is a key working area and four spotlights either side of the beam focus on it.

The position of the spots on either side of the beam is calculated so that even if the spot is adjusted straight down, it will never be visible below the bottom edge of the beam. So as one looks across the room, there is no clutter to destroy the shape of the structure. I always avoid mounting lighting onto the underside of a beam as it makes the spotlight too visible, which is distracting as it then fights with the architecture for attention.

Figure 18.11

By day, the view to the garden from the dining area extends the space.

In the dining area, a wonderful pendant is used, with LEDs integrated to light the glass droplets. Either side, two spots focus on the centre of the table and the surrounding artwork, and a picture light creates a mid-level focus over the fireplace. What is interesting is the huge window looking onto the small garden. During the day, this allows daylight to flood in, but at night, without lighting, the window becomes a vast black void. As soon as the garden is lit, the space is transformed. It's not a suitable size or form for curtains or blinds, so the lit garden adds interest and depth.

Figure 18.12

By night, the space is extended by the garden lighting.

Figure 18.13

Looking in from the garden, the lit house looks wonderful on two levels, the layers of light clearly visible. Two pots right outside the house are strongly lit from above, which immediately draws your eye out when you're inside and then you start looking beyond at the highlighted greenery and water feature.

The small garden at night is lit as an external room using layering techniques. The central water feature has two lights within that provide a ripple effect on the rear wall. The uplit perimeter planting is an essential part of the scheme; small, focused spotlights uplight trees and floodlights show off the lower planting. Two high-level spots on either side of the glazed rear of the building downlight the pots closest to the house to ensure an immediate effect close to the glazing at night.

To control the lights, a preset system operates each area individually and can also control everything together in four scenes. This control can be operated either as a global scene on one of the control plates or on a tablet, or as a double tap of a control plate (if it is programmed in this way). For everyday use, there are four scene settings to each area, so not all the lighting has to be on simultaneously.

These are as follows:
Scene 1. Dull day (Bright)
Scene 2. Early evening (Soft)
Scene 3. Late evening/dining (Mood)
Scene 4. Night (Low)

If a traditional rotary control system had been used, it would have been difficult to get the right balance of light settings quickly, as there would have been multiple (12 or more) rotary knobs in each area to adjust manually.

Each area is set to the same balance of four scenes and it is by combining these that the global scene is achieved.

THE VERDICT

The house is all about providing a flow of light – and harmonising inside with outside, with views from both equally important. It's a very interesting and successful design, and while appearing simple, offers many special areas of well-lit interest.

CHAPTER 19

ALPINE CHALET

THE LIGHTING BRIEF

This contemporary chalet is constructed using traditional timber methods. However, the extensive glazing creates a modern atmosphere where the inside/outside aspect lets one relish the views far more than is possible from a traditional chalet with smaller windows. The lighting needed to be a feature that looked appealing and impressive from both outside and in, and made the balance between the two flow smoothly. Inside, an inviting and luxurious atmosphere is achieved with a combination of striking custom feature pendants and lamps, and many contemporary solutions for a balance of layers of light.

Figure 19.1

This contemporary ski chalet is far more glazed than a traditional chalet, so at night the visible lighting creates an inviting ambience.

THE SOLUTION

Figure 19.2

Square double downlights in the timber ceiling have a more industrial feel than round lights, and are more in style with the interior design.

Figure 19.3

A custom-made installation of suspended glass spheres provides design interest. They are specifically fitted with warm LED light to create an atmosphere of candlelight at night.

In order to maximise the benefits of the glazing and the views at night, the eye needs to be drawn outside. So, miniature spotlights are concealed to uplight the roof structure both inside and out, to create a continued visual connection. On the balcony, miniature 1-watt LED uplights with 10° optics are recessed into the deck to uplight the timber walls and posts. This frames the glazing and brings out the texture of the wood. These architectural lighting effects are subtle yet give the chalet an identity at night. They're also effective at drawing one's eye out past the glazing, rather than just being faced with a reflection of the interior in the glass, which becomes black at night without exterior lighting.

The fabulous entertaining area enjoys a double-height space on one side. The master bedroom is situated on a mezzanine above the seating area, and enjoys the same mountain view.

The open-plan living space has a contemporary fireplace at its centre, with the seating area under the mezzanine. At one end, stairs lead to the master bedroom suite and the dining area is located at the other, double-height, end.

The single-height seating area offers more intimacy. A timber-panelled ceiling adds to the warmth, as do the low-glare contemporary black square directional lights. Their arrangement creates focus on the central red stool using narrow-beam optics. Other directional lights around the perimeter provide infill ambient light with wide-beam optics. They are carefully orientated to ensure they are not directly over any seating, as it is uncomfortable to be under a light and have to shift to avoid glare. Low-level lamps give an intimate layer of soft ambient light around the sofas.

Figure 19.4

In the living area, the wall light on the double-height wall can be manoeuvred to provide reading light on the sofa and reflected ambient light off the timber wall. Directional spotlights bring light to the centre of the room.

The double-height dining area has a sloping ceiling and, rather than a single pendant, the void was filled with an installation of hand-blown glass globes of different sizes which cross the boundaries between light and art. They have a warm colour temperature and are dimmable. The dining area is grounded at each end with mid-level standard lamps, which offer a soft, mid-level layer of light.

The double-height timber wall was a challenge to light. We selected a dramatic wall light that could be manoeuvred to be a task reading light or reflect off the wall, providing soft indirect ambient light.

The stairs leading to the master bedroom are lit with linear LEDs under each tread, making the steps appear to float. Reflected lighting from these also adds to the ambient light in the space.

Figure 19.5

In the master bedroom, an extended headboard houses concealed linear lighting to graze up and highlight the ceiling. Opposite, individual uplights in a raised extended hearth run the length of the room.

The master bedroom headboard is offset from the wall by 40mm to incorporate a warm LED uplight behind. Care is taken to ensure the LED light is spaced at least 25–30mm away from the wall to soften the line of light and create a soft wash of light. (If an LED behind the headboard is positioned too close to the wall, then the light becomes a strong visual line and does not appear to travel up the wall.) On the opposite wall, a raised hearth requires a different solution as there would be a direct view into a linear LED. Instead, a linear LED below the shelf shines soft light onto the floor. On this black raised hearth, miniature low-glare 1-watt uplights with a baffle uplight the wall either side of the fireplace. When dimmed, these appear like a row of nightlights.

Figure 19.6

A decorative pendant (partially in view at the top-centre of this photograph) gives ambience in this bathroom, and wall niches offer linear lighting. Floor spots create a pleasing effect on the bathtub. The suspended mirror in front of a window enjoys bright daylight; by night it is lit with a frosted illuminated frame.

The master bathroom's sloping soffits created a lighting challenge. The ambient light is from a central pendant and the low timber walls feature recessed uplights, centred on each panel. These create reflected light off the sloping soffits and ensure a feeling of maximum width by lighting the extremity of the space. To add detail, a linear LED defines the niche behind the bath, likewise in the shower niche and under the bench. A good facial task light is achieved at night with a frosted backlight around the perimeter of the mirror frame.

Figure 19.7

All bedrooms in the chalet have their own lighting solutions, with downlit cupboard lighting, a selection of stylish lamps and dedicated reading lights offering optimum task light.

In the guest rooms, ambient light comes from four square directional downlights and a soft effect is achieved by concealed linear light to wash down the curtains, together with attractive lamps. Concealed linear lighting in some rooms, integrated behind the bedside tables and also behind built-in drawers, provides a soft, night-time mood. For task lighting, reading lights are placed either side of the bed.

Figure 19.8

Concealed linear lighting at the back of the shower offers soft ambient light and makes the bathroom feel larger; two spots offer a focused light.

Figure 19.9

A halo light around a bathroom mirror reflects off the side walls, making it a good alternative to wall lights.

In this contemporary bathroom, the shower has a linear LED set into a slot, which softly lights the tiles but does not provide enough task light, so two downlights in the shower create focus. A further two downlights emphasise the basin area. The recessed slot created is also a useful device to conceal the extract fan.

In another bathroom, the mirror has a halo light which reflects off the side walls to give a flattering side light to the face. This bounced-back light is an alternative solution to wall lights.

Every luxury chalet has a spa area and this chalet is no exception. One of the key challenges is to create the appropriate mood in the basement with no natural light.

Figure 19.10

This underground spa is lit very effectively. Attention is drawn to the textured walls, which are in harmony with the underground atmosphere, and the mirror is excellent for reflecting light. Battery-operated LED lanterns reflect in the water. Behind the loungers, submerged light sources graze light up the stone wall.

The key element of the lighting is to ensure the ceiling over the pool is free from downlights; this is important not only from a maintenance point of view but also visually. The facing wall is of rough stone that almost feels like one has dug into a mountain. Between the stone are mirrored panels – a clever device to add a sensation of space as one sees another pool in the reflection. The challenge is to light the stone wall, which appears to disappear directly into the water behind the small platforms for the daybeds. The solution is an underwater LED light grazer positioned close to the stone wall and concealed by the daybed platforms. This has a dramatic effect both on the texture of the stone wall and on the timber ceiling.

The low-level lanterns beside the daybeds are battery operated, easily moved and add a low-level layer to the light.

THE VERDICT

This luxury chalet is successfully lit with a variety of appealing decorative lighting and concealed sources, offering warm ambient light that gives a feeling of stylish comfort. The outside and inside lighting balances well, allowing the chalet views to be enjoyed even at dusk and beyond.

CHAPTER 20

CONTEMPORARY VILLA

THE LIGHTING BRIEF

This villa offered many opportunities for a clean and contemporary lighting design. With numerous indoor and outdoor leisure spaces, there is plenty of room for creative lighting solutions. The building design is in a completely natural and neutral material palette, with pristine detailing throughout. Lighting has been interwoven into all the details from the earliest point of the design process to ensure a fully integrated scheme. The gentle tone of the palette offers interest by providing a variety of textures, which can be dramatically illuminated. The spatial intervention of a central courtyard that the property circulates around provides a focus and visual link to the surrounding rooms, which the lighting highlights. As soon as you arrive and are led up the immaculate, palm-lined walkway by the lighting, you feel a sense of what awaits.

Figure 20.1

Domed, cat's-eye-style LEDs skim light across the walkway to this grand villa, leading you up to the main entrance, heralded by four uplit palms.

THE SOLUTION

Figure 20.2

The peaceful central courtyard allows for a breeze through the property and visual circulation. The lighting is simple but effective when viewed from different angles.

Figure 20.3

A set of uplights highlight the timber wall in the internal corridor around the courtyard.

The entrance lighting defines your route and your first impression of the villa. Small LED domed lights throw light across the path, leading you to the front door. Up-and-down wall lights around the entrance and windows emphasise the facade and uplight projecting canopies. The narrow slot windows glow from within, adding to the visual impact.

There are three particularly strong architectural design elements: the simplicity of the circulation around the courtyard, the powerful sculptural staircase that is a key vertical linking element in the building, and the material palette in similar tones but varied textures. Where one area's surfaces might be honed, another will be polished, while in essence being the same material.

On entering the villa, the courtyard dominates the centre, bringing daylight to the mid-point of the deep plan. Circulation around the courtyard is detailed very simply, so at night it almost becomes part of the courtyard area itself. The timber wall is uplit and reflected light provides the ambient light to the corridor. At night this is separated by a glazed wall, so appears to be part of the courtyard when viewed from the opposite side, and there is total inside/outside interaction.

The focus of the courtyard is the tree in its planter, delineated with a concealed linear LED at its base. The tree is uplit from within the planter and downlit from a spotlight from a high level. The perimeter stone wall is uplit in a similar way to the internal timber wall, continuing themes of lighting to emphasise the material palette.

Figure 20.4

The dining areas in the kitchen offer contemporary simplicity. Background lights set the soft ambience, while our eye is drawn to the decorative pendants: filament lamps with warm lights over the kitchen island and circular pendants, whose height is adjustable, over the breakfast table.

Figure 20.5

This sophisticated bar and seating area has two focal points: the bar and the drinks table. Less noticeable, providing ambient light, is a halo of perimeter lighting around the timber ceiling, emerging from its edges. This soft effect continues through the villa.

The family kitchen extends along the rear, with access to the garden and pool area. Here, the back wall concealing the fridges and ovens is lit with an indirect LED strip, providing a soft wall wash effect for general ambient light. Decorative pendants are the focus over the kitchen island. They are an exposed filament lamp; if on full, they would be uncomfortably bright, so they are dimmed. Narrow-beam downlights in between do the task of lighting the kitchen island. Individual LEDs have also been added to the shelves behind.

Three decorative pendant fixtures hover low over the breakfast table, providing a sense of intimacy in this large open area. Introducing decorative lights and suspending them low over a table can help zone an area and make it feel more intimate, despite the scale of the ceiling. By combining the direct effects of downlights, the indirect LED perimeter glow, the joinery lighting and the decorative elements, lighting layers are achieved at all levels, making this large contemporary kitchen inviting whatever the time of day or night.

The adjoining sitting and bar area has a timber ceiling, so black, square, fully directional fixtures seemed the most appropriate solution and work well when integrated into the linear black ventilation slots. Focus is on the bar and coffee table between the seating, with narrow beams, while medium beams provide infill general light. The impressive timber bar area has various elements of light to maximise its importance, from the light under the bar top onto the stools to the integrated combination of linear lights and downlights in the bar itself.

The pale floor reflects light well, and the dark ceiling creates a sense of intimacy in a large space. Directional spotlights from above pool light on the back of the seating, leaving the main seats unlit, which allows for a more comfortable feeling. The lighting halo around the edge of the timber ceiling creates a complete cornice of light, giving the ceiling a floating effect, and also shines down the sheer curtains to give them a lighter feel.

The central staircase is the key statement of the villa, and each corridor leading to it is important. In the passage leading from the basement spa area to the staircase, the uplights define the route on either side, while defined pools of light are like stepping stones leading you through. When lit, the three-dimensional artwork creates unusual shadows that add to the interest of the piece.

Figure 20.6

Lighting creates wonderful drama on this passageway to the spa. The artwork comes to life with spotlighting from above. Along the corridor, intermittent uplights create a linear wash above and pools of downlight along the floor have a combined dramatic effect. The warm light at the end draws you in.

Figure 20.7a
Figure 20.7b

The red pot gives definition and interest to this space, as does the lighting. Note the stone stairs mirrored in timber in the flight just visible above (a). The stairs have added definition when lit (b).

The sculptural central staircase goes between three levels and changes material as it moves from lower ground to first floor. At its base, it is constructed of stone and becomes timber as it ascends through the villa, yet the style and detail of the individual slabs was maintained and lit in the same way.

Figure 20.8

The light installation of glass droplets filling the stair void is inspired by rain (rare in Dubai). The glass drops are lit from above and below.

Both stone and timber treads conceal an LED strip to light each step. At the back of each tread, another LED strip is integrated to uplight the underside of each step, making the stairs a sculpture in the space. The lighting makes each tread appear to float.

Figure 20.9

The consistency of calm, natural materials comes to the fore with lighting and with the simple but beautiful minimalist design.

Contrasting with the strength of the stairs at the centre is an ethereal custom installation of suspended glass droplets that run through the central void. These glass droplets are inspired by rain and, rather than lit individually, they are lit at the bottom with recessed 10° uplights and at the top with a similar arrangement of narrow-beam downlights. The impact of light is that it catches the glass droplets and makes them appear to be internally lit.

The high level of detail is reflected in the consistency of the material palette and the way these materials are used throughout the property.

Figure 20.10a
Figure 20.10b

The juxtaposition between stone and timber seen in the central staircase is again evident in this private study (a). The backlit shelving brings out the texture of the stone, while downlights highlight specific objects (b).

In this study, the background ambience is from the perimeter halo light, with the standing lamp providing a visual focus. The shelves have two lighting solutions: small, shallow downlights for specific objects in the smaller sections and in the larger section the rough stone texture at the back is emphasised with a linear LED concealed behind the polished timber shelves. The lighting accentuates the play of materials. Without lighting the back wall, the detail of the texture may have been missed.

In the bathrooms, soft general light is achieved by floating the ceiling off the wall and by backlighting the mirrors. These effects enhance the stone and give a soft glow to the vanity surface below. In addition, decorative pendants and wall lights create a soft forward light that is essential to achieve a good facial light. Downlights are added to emphasise the stone vanity top and ensure the stone floor and all surfaces are celebrated.

The steam room has a darker stone than the bathrooms, which is polished in the steam room itself and textured in the changing area. The contrast of the effects is emphasised by light, and the play of treating the same stone differently makes the space seem harmonious and calm. Using a simple palette of materials and tones yet in different finishes – textured, honed and polished – creates the homogeneous consistency that is felt throughout the villa.

The children's bedrooms all have a sense of fun. The high ceiling has allowed a mezzanine to be created above the bed, accessed via steps lit with spots and exited via a slide, lit with a smooth linear light (see Figure 12.6). Soft indirect light shines down the curtains and behind the headboard. The shelves on the mezzanine and either side of the bed are also lit, adding mid-level layers of light that decorative lamps would provide in a more traditional bedroom.

The shelving nooks are alternatively lit so books will be unlit and objects lit, also providing a fun, chequered effect. Downlights focus on the table as a practical task light.

The gym finishes are dark and dramatic, as there is little reflected light. There is a softness from the linear wash onto the timber wall and this contrasts with the strong directional light on the gym equipment. These square double fixtures have been integrated into the black slots of the air conditioning. As there is no outside view, having a glazed screen to the garage increases the sense of space and allows the owner to see their prized car collection.

Figure 20.11

This gym poses a challenge as multiple dark finishes make it tricky to bring brightness. Energy is created from spotlights on the equipment.

Figure 20.12

Outside entertaining in the evening can be enjoyed year-round in the warm temperatures of Dubai, so this seating area is glamorously lit with strong linear LEDs around the base and a halo-lit art feature on the wall. The palms are also uplit in the background.

Lighting the exterior landscape well is essential, as darkness falls between 6 and 7pm in Dubai, yet the space is used almost entirely in the evening as it's too hot during the day. Pergolas provide shade in the day but also offer a sense of containment at night. Spots are incorporated within the timber slots to act as functional light. The fixed seating and tables have light built under them that provides a soft reflected light off the pale stone. A focus is created by the halo-lit art piece and, in the background, uplight to the palms helps to create the mood.

THE VERDICT

This home offers a calming and luxurious coolness, with soft-toned natural materials throughout, successfully highlighted with light. The exterior and interior are in harmony and the entertainment spaces feature strongly, with more dramatic lighting. But equally, flow spaces are important – the stairs and passageways – and we have created interest in them using multiple layers of light.

CHAPTER 21

DUPLEX PENTHOUSE

THE LIGHTING BRIEF

This duplex penthouse apartment is a work of art, with immense attention to detail and a strong material palette of blackened steel, oak and artistic finishes, such as honeycomb-glazed screens. The combination of materials creates a beautiful backdrop for the textured furnishing scheme that softens the strength of the steel and oak. Lighting this atmospheric home was a creative adventure. Working with strong materials allowed experimentation and some striking light effects, such as the light filtering through the honeycomb glass. This city-centre penthouse also features an exceptional winter garden, which can be enclosed or completely opened out to a terrace, allowing for an entertainment space with a selection of contemporary lighting options.

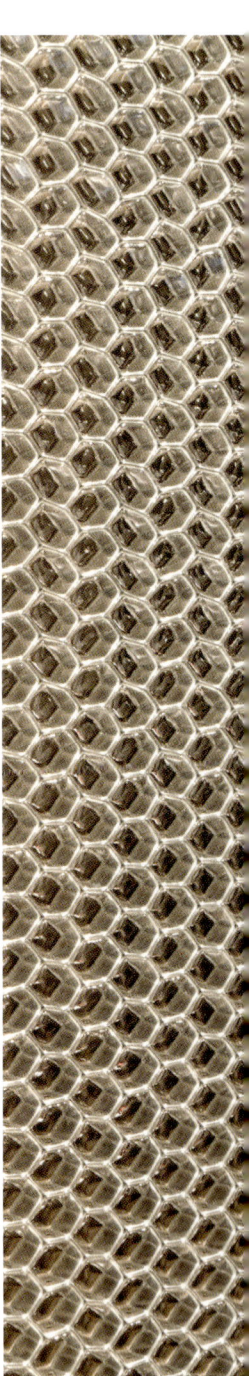

Figure 21.1

Honeycomb-glazed double doors catch natural light during the day and sparkle beneath directional spots at night.

THE SOLUTION

Figure 21.2

The oak stairs are set against a blackened-steel wall; recessed lights in the string illuminate each step and provide a glow on the opposite side. There is also a soft pool of downlight at the base of the stairs.

Figure 21.3

The dark bedroom walls absorb light. The main ambience is from downlights reflecting off the end of the bed, bringing light to the centre of the room. Warmth from the lit timber in the bedroom lobby and bathroom draw one in.

The duplex is linked by the central staircase, set against a blackened-steel wall which is lit from the inside string of the stairs. Lighting falls across the oak treads and reflects off the steel wall.

The master suite is on the lower level and the main living area is on the floor above, incorporating a dining area, a kitchen and relaxed seating. There is a sloping glazed roof over the galleried dining area, linking the two floors.

The master suite invites one in through a dressing lobby at the base of the stairs.

The dark linen walls of the bedroom create an intense and moody atmosphere, with pools of downlights on the bed adding to the drama and providing a focus to the centre of the room, which without it would feel too dark. The general light in the room comes from trimless directional square downlights, softened by table lamps. With these dark finishes, the room is never bright but has a wonderful atmospheric quality. For reading, there is a wall-mounted task light each side of the bed.

Figure 21.4

The timber-clad wall in the master bathroom is lit with an indirect detail in a slot that reflects off the soffit. Uplights, downlights and wall lights ensure a well-lit vanity area. Accent to the timber bath and the volcanic tiles comes from narrow-beam downlights.

Leading on from the bedroom is the dressing room and master bathroom. Both are mostly panelled in oak, which provides a soft, warm glow from the reflected light off the surface of the wood. The bathroom continues with the oak finishes. Volcanic black tiles and a dark patinated mirror replace the steel in this area. Lighting concealed in slots grazes down the timber walls and the black square trim fixtures add accent light.

The dressing-room lobby links the bedroom to the main dressing room, which also links to the bathroom. At low level in the lobby are two immaculate rows of shoes lit with a continuous warm 2,400K LED strip. Higher up, another 2,400K LED strip grazes up the patinated mirror, giving a soft ambient reflected light off the ceiling. In contrast, a direct pool of light comes from black, square LED downlights with medium-beam optics. These have been selected to give an urban, industrial feel to match the blackened-steel detailing used throughout the property. These pools of light lead up to the main dressing area.

The lamp on the side joinery unit is the visual focus that attracts one's attention as the main light source, with its soft ambient lighting. It adds another dimension when combined with the LED strips on the shoes and the uplights and the downlights. It is this layering of light throughout each room that makes the apartment feel so comfortable. Each effect is controlled individually so that the light is balanced to suit the scene for the time of day as follows:

Scene 1: Morning (Bright)
Scene 2: Early evening (Soft)
Scene 3: Late evening (Mood)
Scene 4: Night (Low)

Figure 21.5

This small lobby, linking the dressing room with the master bedroom, illustrates successful layers of light to build up the whole. The focus is the lamp, but general light is from multiple other sources: recessed downlights, an LED strip uplighting the mottled glazing that separates the bathroom, and low-level accent light to the shoes.

Figure 21.6

Lighting within the glazed cabinets in the dressing room creates added depth and warmth.

Figure 21.7

The magnificent top-floor combines a kitchen, living area, dining area and roof-top conservatory. Individual stair lights draw one up. The dining area to the left is challenging as it's under a sloping glazed ceiling. The solution is custom pendants with a bracket designed to fit the junction of the solid ceiling and glazing.

Once in the dressing room, the room feels large, as all the cupboards are glazed and internally lit with concealed vertical LED strips. Vertical strips allow more flexibility of moving shelves up and down.

Square, black, trimless downlights are used to focus light onto the cupboards for additional task light when clothes are removed, helping to differentiate between black and navy, for example.

The downlights are also used as task light over the central island. In the centre of the island, a structural column in blackened steel supports two bracket lights with parchment lamp shades. These introduce another layer and visual focus, in contrast to the strong design elements, with their soft infill light, which is more flattering than downlights. The full-length mirror to the left of the image has two forms of light: one a soft, decorative, shaded light above, the other two LED strips integrated into the mirror behind frosted glass to provide a head-on, shadow-free light for the person standing in front of the mirror. This is locally controlled to provide the perfect effect when looking in the mirror.

The stairs to the main open-plan living area are individually lit using small, square directional lights on each tread, built into the left-hand string. They highlight the timber tread and give a soft glow on the blackened-steel panels opposite. The staircase divides the dining zone from the living area and ahead is the kitchen.

Figure 21.8

In the winter garden, striking copper panels are lit with wall-mounted spotlights to send shafts of light down. A suspended light over the round table offers task and ambient light; a downlight was added to its base to give a strong pool of light on the table below.

The dining table is in the galleried balcony area. It is below the large, glazed, sloping rooflight that links to the floor below, so lighting for the table was a challenge. The solution is a custom-made pendant fixed at the edge of the solid ceiling and rooflight, incorporating a general light through the side panels, giving a glow to the ceiling. On a separate circuit, three recessed lights are integrated below the pendants to provide focused light on the table itself. To get the best effect, both elements are controlled independently.

The dazzling rooftop space off the upper living area is known as the winter garden, and comes into its own with the glass doors fully open onto the terrace.

The copper panels have a wall-mounted light to emphasise the unusual colour and texture. Between the skylights, a square downlight infills the centre of the space. For softness and warmth and a visual focus of light, a table lamp is positioned at the end of the sofa. A feature pendant over the table has two light sources incorporated: LED filament lamps that glow through the custom woven shade, and also a focused downlight set into the base of the shade for focus on the table. The lit planting on the terrace also adds to the inside/outside feel and extension of space in this narrow room.

Looking back past the sitting room is the kitchen and relaxed dining area. The lamps in the sitting room provide the ambient lighting either side of the sofa and TV unit. Downlights are divided into separate control channels, one to focus over the ottoman, drawing people into the seating area. Around the perimeter, square double downlights with a wider beam add to the general light and are also used to highlight the unit below the TV.

In this space, the kitchen/dining area is very inviting. A timber slatted screen can slide around the curved bench to enclose it and conceal the windows to the right, acting as a shutter filtering the light during the day and enclosing the space at night. The pendant is like the one in the winter garden, with two light sources, and the quirky wall light gives a focused light for reading. Uplights behind the bench to light the timber screens and ceiling above have a transformative effect.

To achieve the uplight behind the bench, a linear strip would not have had enough impact, so multiple 1-watt 10° uplights with a spreader lens are built into the plinth behind the bench. This highlights the curved form of the timber screen. The miniature uplights are set approximately 200mm below the top of the bench and approximately 200mm apart, to ensure the beams overlap each other to make a continuous band of light.

Figure 21.9

In the sitting room, lamps and dimmed perimeter downlights create ambience; however the focus is on the central ottoman, lit with greater intensity. The end view of the kitchen is fundamental to the overall impact of the room.

Figure 21.10

In the kitchen, pendants provide the visual lighting focus and style. General light comes from the double downlights, layered with soft uplight to the curved timber wall, plus integrated downlights and backlights in the shelving. Each adds a different effect and combined bring the perfect balance.

Finally, in the kitchen, a decorative element is introduced on the central blackened-steel column supporting two cone-shaped, rise-and-fall pendants over the island. This is a strong design feature, while general downlights do the work. The open shelving is lit with a combination of mini downlights through the glass shelves and an LED-strip backlight, adding another layer of light.

The control of lighting is important. A preset system is used here as this is particularly useful in open-plan areas, as each zone is considered as an individual room and has four scenes: bright, soft, mood and TV or low. There are keypads in each local area operating each zone as an individual room and by a simple double tap a global scene can come on, meaning the entire open-plan area could all light up at a set level to ensure a totally balanced lighting effect. If one then wanted more task lighting in the kitchen, this could be raised locally. A double tap off means the whole area can be switched off simultaneously.

Setting lighting scenes can also be done on a tablet or phone. Preset lighting for open-plan areas always makes the lighting easier to operate, avoiding multiple switches to set the mood.

THE VERDICT

In all projects, lighting should be an integral part of the design from the outset. In this one, it was. Decorative lighting was an important focus for the design in order that the multiple architectural lighting effects work, were never obvious. While doing the majority of the work. This layering is critical throughout. Attention to detail and close collaboration with the designers resulted in the ultimate success of this design.

What is essential in this penthouse scheme is that the exceptional interior finishes are enhanced with light, and the atmosphere is inviting without it being obvious how many layers of light have been used to ensure the perfect overall result.

GLOSSARY OF TERMS

Accent light – Lighting which highlights a particular place or object. A picture light provides accent lighting to an artwork.

Ambient light – The general lighting in a room, which can come from various sources.

ANSI Chromaticity Standard – ANSI stands for American National Standards Institute. The ANSI chromaticity standard is a communication tool used between LED makers and users, standardising the white colour variation for indoor lighting.

Baffle – A device used to deflect, check or regulate flow of light. It usually refers to the trim found on recessed lighting fixtures to soften the glare of light.

Barrisol – A Barrisol is a stretched PVC sheet used as a ceiling, or as an area of a ceiling. The material used allows the uniform diffusion of the light source placed behind – a Barrisol Lumière appears almost as a ceiling light box.

CCT – This stands for correlated colour temperature. This means the colour temperature of a white LED. CCT is defined in degrees kelvin, where a warm light is around 2,700K moving to neutral white at around 4,000K to cool white, at 5,000K or more.

Chamfered reveal – A niche with sloped, cut-away edges, often housing a window.

Chromaticity – The quality of colour, independent of brightness.

Close offset – Offsetting lighting near to an edge.

COB – Stands for chip on board. In the LED market, chip on board refers to multiple LED chips (typically nine or more) bonded directly to a substrate by the manufacturer to form a single module.

Coffer – Coffer literally means indent. Usually used in terms of a ceiling coffer, it is a sunken panel in a ceiling.

Colour temperature – The colour temperature of a white LED defines the warmth of the light. (See CCT, above.)

Cove – The ceiling cove is the area between the ceiling and wall, and a coved ceiling is usually curved or decoratively joined to the wall. This decorative element can conceal lighting.

CRI – This stands for Colour Rendering Index. It is the measurement of how colours look under a light source when compared with sunlight.

DALI dimming – DALI stands for Digital Addressable Lighting Interface. They are dimmer switches which use a digital protocol to send a dimming control signal to a driver down separate wires to the mains supply.

Diode – A two-terminal electronic component that conducts current primarily in one direction; it has low resistance in one direction and high resistance in the other. The diode can be viewed as an electronic version of a check valve.

Dolly switch – A more old-fashioned switch style, using an up-down lever, rather than a push button or rotating dimmer.

Downlights – Lights which project light downwards.

Downstand – A piece of joinery attached to something and hanging down, for example under kitchen cabinets or in shelving, to hide the light fixtures.

Elevation – An architectural term: an elevation is the front, back or side of a building, represented in 2D on a plan.

Extrusion – In lighting, an extrusion is the profile casing for an LED light, specifically created to fit the shape of the LED light and enable a designer to easily place LED lights exactly where they are required without the need of more traditional light fixtures.

Fibre optics – Thin, flexible fibres with a glass core through which light can be sent. Ideally should not exceed 10m.

Floor washers – Floor washers (or floor washing) are low-level lights which skim light across a floor (or steps).

Glare guard – An accessory or baffle used to conceal a light source.

Gobo – A slide that is specially cut to shape, used to frame pictures or sculptures exactly, with no light spill.

Graphic line – A sharp line of light created by a linear LED.

Grazing optics/grazer – This optic changes the light from a narrow to an elongated beam, which is used close to a wall to light it and to emphasise the texture.

Halogen – Halogen lights were used before LEDs, and are lamps and radiant heat sources using a filament surrounded by the vapour of iodine or another halogen.

Heatsink – A heatsink is a passive heat exchanger that transfers the heat generated by an electronic or mechanical device. For LEDs this is usually a solid metal extrusion which dissipates heat away from the LED.

Honeycomb louvre – A honeycomb-shaped louvre is used to prevent glare from the lamp.

IP rated – The IP Code, or Ingress Protection Code, sometimes referred to as International Protection Code, classifies and rates the degree of protection provided by mechanical casings and electrical enclosures against intrusion, dust, accidental contact and, crucially for lighting bathrooms and swimming pools, water.

Joinery – The wooden components of a building, such as stairs, doors, fitted shelving, some furniture and door and window frames, viewed collectively.

Kelvin – A unit of temperature. A kelvin is the base unit of temperature in the International System of Units (SI), having the unit symbol K.

Lamp – In technical usage, a replaceable component that produces light from electricity is called a lamp. However, the misunderstanding comes as lamps are commonly known as light bulbs.

Leading-edge dimmers – Originally used for dimming incandescent and halogen lamps. Due to their original use, these dimmers have a high minimum load, making them less useful for low-energy lighting such as LEDs.

LED – Stands for light-emitting diode. A light-emitting diode is a semiconductor light source that emits light when current flows through it. Electrons in the semiconductor recombine with electron holes, releasing energy in the form of photons.

LED driver – A device that regulates the current and voltage to the amount required for an LED, or an array of LEDs, to operate.

Linear LED – An LED lighting source in a straight line using multiple LEDS on a tape and often located in a extrusion with a diffuser.

Louvre – Grid type of optical assembly used to control light distribution from a fixture.

Lumen – A unit of light flow: the lumen rating of a lamp is a measure of the total light output of the lamp.

Luminaire – A complete electric light unit or fixture.

Luminance – Measure of brightness perceivable to the human eye, in terms of luminous intensity per unit area of light shining on a particular place (object or surface). It's a measure of the amount of light that passes through, is emitted from or is reflected from a particular area, and falls within a given solid angle.

Lux – Measures luminous flux per unit area. Over time, lux levels will get lower with any type of lighting, although LED lighting is very resilient.

MacAdam ellipse – A MacAdam ellipse is a region on a chromaticity diagram which contains all colours which are indistinguishable to the average human eye, from the colour at the centre of the ellipse. The contour of the ellipse represents the noticeable differences of chromaticity.

Night-light – A low-energy light providing just enough light to navigate at night.

Opal diffuser – A milky, opalescent coating to diffuse the light.

Optic – An important element of an LED luminaire. The optic shapes, focuses and mixes the light created by the LED light source into a shape, for example a wide flood light, a narrow spot or an elliptical wall grazer.

Perimeter ambient light/perimeter wash light – Concealed LED indirect light in a ceiling slot around the perimeter of the room, lighting the walls.

Pilaster – A rectangular column, especially one projecting from a wall.

PIR – PIR stands for passive infrared, and is used with reference to lighting that detects heat, such as that of a person. For example, PIR presence-detector lights switch on when a person is nearby.

Riser – The vertical part of a stair, between and usually supporting each tread. Open stairs do not have a riser.

Soffit – The underside of an architectural structure, such as overhanging eaves, an arch or a balcony.

String – A stair string (or stringer, or stringer board) is the housing on either side of the steps of a staircase, into which the treads and risers are fixed.

Task light – Lighting which highlights an area used for a task, for example reading, preparing food, cooking, working.

TM-30 – A method to measure a light source's light quality.

Torchère – The ornamented stand of a wall-mounted holder for a candlestick.

Trailing-edge dimmers – These more modern dimmers have many benefits over the leading-edge type. These improvements include smoother dimming with less buzzing and interference. Trailing-edge dimmers have a much lower minimum load than leading-edge dimmers, making them far more suitable for powering LEDs.

Treads – The flat part of the staircase on which you tread.

Tungsten – Tungsten is a chemical element, used in old-fashioned lightbulbs that create a warm light. These lightbulbs feature a tungsten filament housed within an inert gas, and when a current is passed through the filament, the naturally high resistance of tungsten causes the filament to glow and emit a warm orange light.

Uplights – Lights which project light upwards.

Upstand – A piece of upstanding joinery, for example above kitchen cabinets or shelving, to hide the light fixtures.

UV – Ultraviolet radiation; UV is a part of fluorescent lighting.

Vanity – Short for vanity unit, which is the joinery that houses the basin in a bathroom.

Wall washing – A technique whereby light is directed to the wall and ambience is created by the reflected light.

INDEX

A

accent lighting 14, 27–9, 35, 88–9, 128
ambient lighting 14, 18–26, 86, 87, 108
ANSI Chromaticity Standard 38
architectural features and materials 7
art works 69, 154, 156, 167, 172, 173, 190
  (*see also* pictures)
attics 132–4

B

baffles 25
barrisols 69
bars 147, 189
basements 136–9 (*see also* leisure rooms)
bathrooms 45, 120–9, 133, 136–7
  case studies 158, 166, 182, 184, 193,
    199
bedrooms 110–19
  case studies 158, 159, 165, 182–3, 194,
    198
Bluetooth Low-Energy Control (BLE) 47
bookcases/shelves 88, 164, 172
bowling alleys 148, 149

C

case studies
  alpine chalet 178–85
  barn conversion 168–77
  contemporary Dubai villa 186–95
  contemporary terraced house 152–9
  duplex penthouse 196–203
  traditional terraced house 160–7
cinema rooms 146–7
coffer lighting 19–20, 66, 102
colour binning 38
colour consistency 38
colour fidelity (RF) 37
colour gammet (Rg) 37
colour quality *27*, 35–7, 55
colour rendering 27, 35–7, 55
colour rendering index (CRI) 36
colour temperature 32–5
controls *see* lighting controls
corridors 6, 12, 65, 66, 69
  case studies 165, 173, 188, 190
courtyards *22*, 142, 168–9, 173, 188
cove lighting 19–20, 21, 34, 100–1, 104, 190
curtains 67, 112, 114, 117

D

DALI (Digital Addressable Lighting
    Interface) 46
daylight 32, *33*
decorative lighting 26, 77–8, 86
dimming 33, 41, 46–7
dining rooms/areas 92–9, 103, 139
  case studies 174, 175, 181, 189, 201, 202
double-height spaces 63, 66, 87
  case studies 153, 155, 180–1
downlighting 25
dressing rooms 118–19, 166, 199–200

E

electrical wiring 5
entrance halls 44, 65–9
entrances 58–63, 187–8
exterior lighting *see* outside lighting;
    transitional lighting

F

facades 62–3
feature lighting 7, 27, 108, 123–5
fireplaces 84, 85
flowers 28
front doors 60–2
fruit bowls 35

G

garden lighting 8, 91, 177 (*see also*
    courtyards)
glare control 24, 73, 89
glass blocks 4, 5
glazing 8, 109 (*see also* transitional
    lighting; windows)
grazing 22, 62, 76, 86, 124, 182, 185
green walls 8
gyms 194

H

handrails 75, 81
highlighting 7, 10–11 (*see also* feature
    lighting)
honeycomb louvres 89

I

ingress protection (IP rating) 122

K

kitchens 18, 23, 24, 29, 55, 100–9
  basement 138
  case studies 155, 163, 174, 203
  understairs 135

L

lampshades 86
landings 73
layering 13
leading-edge dimmers 46
LEDs (light-emitting diodes) 30–9
  colour consistency 38
  colour quality 35–7, 55
  colour temperature 32–5
  colour-changing 39
  colour-tuneable 33, 35
  dimming 33
  drivers 39
  linear 75–6
  optical control 38
  thermal management 38
leisure rooms 140–9
lighting controls 41–7, 50
  bathrooms 122
  bedrooms 114
  dimming 33, 41, 46–7
  dining rooms 99
  living rooms 86, 90
  outside lighting 64, 177
  scene-setting systems 43–5, 177, 199,
    203
lighting levels 50
lighting plans 48–55
lighting symbols 50
linear lighting 75–6
living rooms 82–91
luminance 13

M

MacAdam ellipse 38
mirror lighting 126–7

N
natural light 32, 33
night-lights 133

O
offices *see* studies/study areas
opal diffusers 19, 75, 106, 127, 158
open plan 9
open-plan rooms 53, 171, 200, 203
optical control 38
outside lighting 8, 64, 91, 156, 177, 195 (*see also* courtyards; entrances)

P
perimeter lighting 21, 97
pictures 22, 24, 28, 68, 83, 175
PIRs (passive infrared presence detectors) 47
pool rooms 140–5, 185
pool-house 52
presence detection 47

R
reading lights 90, 112, 114
remote control 44

S
scene-setting systems 43–5, 177, 199, 203
shelving 84, 97, 105, 106, 108, 131 (*see also* bookcases/shelves)
    case studies 162, 174, 189, 193, 203
shower lighting 121, 122, 124
skylights 71, 79, 83
spaces, defining 9
specifications 51
spotlighting 16–17, 28
staircases 15, 70–81
    case studies 157, 170, 181, 190–1, 198, 200
steps 63
studies/study areas 134, 173
swimming pools *see* pool rooms; pool-house

T
task lighting 14, 29, 89, 90, 98 (*see also* reading lights)
    bathrooms 126–7
    kitchens 100, 102, 103, 105–7
texture 7, 13
thermal management 38
time clocks 47
TM-30 colour fidelity index 37
transitional lighting 8, 52, 54, 91, 109, 156, 175, 180

U
under-stair spaces 135, 154
uplighting 18

V
vaulted areas 136–8
volume (space) 6

W
wall grazing *see* grazing
wall washing 3, 7, 21, 23, 24
wardrobes 113, 118–19
WCs 122, 129, 136
windows 8, 29, 54, 62, 133, 156, 175 (*see also* skylights)
wine cellars 139
wireless control 44, 47
wiring 5

IMAGE CREDITS

Lighting design throughout all properties is either by Lighting Design International or John Cullen Lighting.

Figure 0.1: Architect: ANARCHITECT, Photographer: Ieva Saudargaite

Part One Opener: Interior Designer: Taylor Howes Interiors, Photographer: Tom Sullam

Figure 1.1: The Edwardian Manchester Interior Designer: Edwardian In-House Team, Photographer: Andrew Beasley

Figure 1.2 & 1.3: Deirdre Dyson Showroom, Architects: Timothy Hatton Architects, Photographer: Andrew Beasley

Figure 1.4: Interior Designer: Todhunter Earle Interiors, Photographer: James Balston

Figure 1.5: Photographer: James Cameron

Figure 1.6: Richstone Properties

Figure 1.7a-c: John Cullen Lighting

Figure 1.8: Architect: Oliver Morgan Architects, Photographer: Andrew Beasley

Figure 1.9: Design: Chamber Furniture, Photographer: James Balston

Figure 2.1: Interior Designer: Designers Guild, Photographer: Breed Media

Figure 2.2a-c: Architect: ANARCHITECT, Photographer: Ieva Saudargaite

Figure 2.3: Architect: ANARCHITECT, Photographer: Chris Goldstraw

Figure 3.1: Hebanon Fratelli Basile

Figure 3.2: Interior Designer: Juliette Byrne Interiors, Photographer: James Balston

Figure 3.3: Carmody Groarke, Photographer: Christian Richters

Figure 3.4a-b: Architect/Designer: TLA Studio, Interior Designer: Jean-Louis Denoit, Photographer: James Balston

Figure 3.5: Interior Designer: Michaelis Boyd, Photographer: Gavriil Papadiotis

Figure 3.6: Architect: Studio Indigo, Interior Designer: Todhunter Earle Interiors

Figure 3.7: Kitchen Designer: Alistair Fleming, Photographer: James Balston

Figure 3.8: Design: Chamber Furniture, Photographer: James Balston

Figure 3.9: 207 -2011 Old Street Office London, Client: Helical Bar, Architects: Allford Hall Monaghan Morris, Photographer: Andrew Beasley

Figure 3.10: Designer: Yahya, Photographer: Dan Kullberg

Figure 3.11a-b: Interior Designer: Studio Reed, Photographer: Nick Kontou

Figure 3.12: Interior Designer: Designers Guild, Photographer: Breed Media

Figure 3.13: Architects: Franklin Associates Architects, Kitchen Designer: Bulthaup, Photographer: Andrew Beasley

Figure 4.1: Carmody Groarke, Photographer: Christian Richters

Figure 4.2: Zumtobel Group

Figure 4.3 & 4.4: Interior Designer: Juliette Byrne Interiors, Photographer: James Balston

Figure 4.5a-b: Africa Studio/Shutterstock.com

Figure 4.6 & 4.7: Forge

Figure 5.1: Interior Designer: Woolf Interior Architecture and Design, Photographer: Ingrid Rasmussen

Figure 5.2 & 5.3: Interior Designer: Michaelis Boyd, Photographer: Timothy Evan-Cook

Figure 5.4a-b: Photographer: James Balston

Figure 5.5a-c: John Cullen Lighting, Photographer: James Cameron

Figure 6.1: Architect: Stephen Marshall Architects, Interior Designer: Designers Guild, Photographer: Breed Media

Figure 6.4, 6.5, 6.6, 6.7 & 6.8a-b: Interior Designer: Designers Guild, Photographer: Breed Media

Figure 6.2 & 6.3: John Cullen Lighting

Part Two Opener: Interior Designer: Louise Bradley Interiors, Photographer: Ray Main

Figure 7.1: Interior Designer: KNA Design, Photographer: Robert Miller

Figure 7.2: Interior Designer: Juliette Byrne, Photographer: James Cameron

Figure 7.3: Photographer: James Balston

Figure 7.4: Landscape Designer: Osada Design, Architect: BLDA Architects, Interior Designer: Todhunter Earle Interiors, Photographer: James Balston

Figure 7.5: Interior Designer: Juliette Byrne, Photographer: James Balston

Figure 7.6: Landscape Designer: Osada Design, Architect: BLDA Architects, Interior Designer: Todhunter Earle Interiors, Photographer: Jeff Brown

Figure 7.7: Architect: Jamie Falla Architecture, Photographer: Richard Brine

Figure 7.8: Architect: Richard Wagner, Interior Designer: XBD Collective, Photographer: Aasiya Jagadeesh

Figure 7.9: Interior Designer: KNA Design, Photographer: Robert Miller

Figure 7.10: John Cullen Lighting Paris Showroom

Figure 7.11: Architect: Studio Indigo Interior, Interior Designer: Todhunter Earle Interiors, Photographer: Luke White

Figure 7.12: Architect: Thomas Croft Architects, Interior Designer: Francis Sultana

Figure 8.1: Interior Designer: Studio Indigo, Photographer: Andrew Beasley

Figure 8.2a: Interior Designer: Designers Guild, Photographer: Breed Media

Figure 8.3: Interior Designer: Karen Howes

Figure 8.4: Interior Designer: Louise Bradley Interiors, Photographer: Ray Main

Figure 8.5: Architect: BLDA Architect, Interior Designer: Todhunter Earle Interiors, Photographer: Jeff Brown

Figure 8.6: Architect: AndArchitects, Photographer: Richard Lawson

Figure 8.7: Design Team: Studio Indigo, Todhunter Earle Interiors

Figure 8.8: Interior Designer: Finchatton, Photographer: Richard Waite

Figure 8.9: Harrison Sutton Partnership, Photographer: Richard Lawson

Figure 8.10: Architect: ABA International Ltd

Figure 8.11: Deirdre Dyson Showroom, Architects: Timothy Hatton Architects, Photographer: Andrew Beasley

Figure 9.1: Architect: Studio Indigo, Interior Designer: Todhunter Earle Interiors, Photographer: Luke White

Figure 9.2: Interior Designer: Alexandra Dixon Interiors, Photographer: Luke Foreman

Figure 9.3: Interior Designer: Louise Bradley Interiors, Photographer: Ray Main

Figure 9.4: Architect: Timothy Hatton Architects, Interior Designer: Sybille de Margerie, Photographer: Fabrice Rambert

Figure 9.5: Architect: BLDA Architect, Interior Designer: Todhunter Earle Interiors, Photographer: Jeff Brown

Figure 9.6: Architect: Studio Indigo Interior, Interior Designer: Todhunter Earle Interiors, Photographer: Luke White

Figure 9.7: Interior Designer: Hollie Bowden Interiors, Photographer: James Balston

Figure 9.8: Interior Designer: Michaelis Boyd, Photographer: Gavriil Papadiotis

Figure 9.9: Interior Designer: Designers Guild, Photographer: Breed Media

Figure 10.1: Interior Designer: Louise Jones Interiors, Photographer: Andreas Von Einsidedel

Figure 10.2: Interior Designer: Michaelis Boyd, Photographer: Gavriil Papadiotis

Figure 10.3a-b: Architect: TLA Architects, Interior Designer: Jean-Louis Denoit, Photographer: James Balston

Figure 10.4: Design: Chamber Furniture, Photographer: James Balston

Figure 10.5: Treehouse Hotel, London, Photographer: Eric Laignel

Figure 10.6: Interior Designer: Brian Woulfe, Photographer: Jonathan Bond Photography

Figure 10.7: Interior Designer: Todhunter Earle Interiors, Photographer: Ray Main

Figure 10.8a-b: Interior Designer: Michaelis Boyd, Photographer: Timothy Evan-Cook

Figure 11.1: Architects: TLA Architects, Interior Designer: Jean-Louis Denoit, Photographer: James Balston

Figure 11.2: Interior Designer: Taylor Howes Interiors, Photographer: David Garcia

Figure 11.3: Interior Designer: Todhunter Earle Interiors, Photographer: James Balston

Figure 11.4: Architect: Ben Wood of Bertram Design, Photographer: James Balston

Figure 11.5: Architect: Ian Adam Smith Architects, Photographer: James Balston

Figure 11.6 & 11.9: Kitchen Designer: Alistair Fleming, Photographer: James Balston

Figure 11.7: Photographer: Lucy Butler-Walters, Joinery: Barr Joinery

Figure 11.8: Interior Designer: Alison Henry, Photographer: James Balston

Figure 11.10: Designer: Eggersman Kitchen, Photographer: James Balston

Figure 11.11: Garden Design: Hay Joung Hwang, Landscape Designer: Randle Siddeley, Photographer: Georgina Viney

Figure 12.1: Interior Designer: Studio Reed, Photographer: Nick Kontou

Figure 12.2a-b: Architect: TLA Architects, Interior Designer: Jean-Louis Denoit, Photographer: James Balston

Figure 12.3a-b: Interior Designer: Brian Woulfe, Photographer: Jonathan Bond Photography

Figure 12.4: Treehouse Hotel, London, Photographer: Eric Laignel

Figure 12.5: Interior Designer: Sophie Paterson Interiors, Photographer: Ray Main

Figure 12.6: Architect: ANARCHITECT, Photographer: Chris Goldstraw

Figure 12.7: Architect: BLDA Architect, Interior Designer: Todhunter Earle Interiors, Photographer: James Balston

Figure 12.8: Regent's Crescent, Interior Designer: Millier

Figure 13.1: Interior Designer: Studio Indigo, Photographer: Andrew Beasley

Figure 13.2: John Cullen Lighting

Figure 13.3: Interior Designer: Designer's Guild, Photographer: James Balston

Figure 13.4: Interior Designer: Juliette Byrne, Photographer: James Balston

Figure 13.5: Interior Designer: Michaelis Boyd, Photographer: Gavriil Papadiotis

Figure 13.6: Interior Designer: Todhunter Earle Interiors, Architect: Smallwoods

Figure 13.7: Interior Designer: Louise Bradley Interiors, Photographer: Ray Main

Figure 13.8: Architect/Interior Designer: Stefano Dufour - DUMAA Architects

Figure 13.9: Regent's Crescent, Interior Designer: Millier

Figure 13.10: Allbright, Mayfair, Interior Designer: Suzy Hoodless, Photographer: Taran Wilkhu

Figure 14.1 & 14.2: Interior Designer: Louise Bradley Interiors, Photographer: Ray Main

Figure 14.4: Interior Designer: Juliette Byrne Interiors, Photographer: James Cameron

Figure 14.5: Interior Designer: Taylor Howes Interiors, Photography: Luke White

Figure 14.6: Harrison Sutton Partnership, Photographer: Richard Lawson

Figure 14.7: Interior Designer: Alison Henry, Photographer: James Balston

Figure 14.8: Deirdre Dyson Showroom, Architects: Timothy Hatton Architects, Photographer: Andrew Beasley

Figure 14.9: Interior Designer: Louise Bradley Interiors, Photographer: Ray Main

Figure 14.10: Architect: Thomas Croft Architects, Interior Designer: Francis Sultana

Figure 14.11: Richstone Properties, Photography: James Balston

Figure 14.12: Templeton House, Richstone Properties

Figure 15.1: Interior Designer: Taylor Howes Interiors

Figure 15.2: Architect: Timothy Hatton Architects, Interior Designer: Sybille de Margerie, Photographer: Fabrice Rambert

Figure 15.3: Interior Designer: Studio Indigo, Photographer: Andrew Beasley

Figure 15.4: Carmody Groarke, Photographer: Christian Richters

Figure 15.5: The Lanesborough Club & Spa, Interior Designer: 1508 London

Figure 15.6: Architect: Ian Adam Smith Architects, Interior Designer: Gosling Ltd, Photographer: James Cameron

Figure 15.7 & 15.8: Heckfield Place Hampshire, Architect: Sprately & Partners, Interior Designer: BWT Interiors

Figure 15.9 & 15.10: Ham Yard - Firmdale Hotels by Kit Kemp, Photographer: Simon Brown

Part Three Opener: Landscape Designer: Marcus Barnett, Photographer: Luke White

Figure 16.1, 16.2, 16.3, 16.4, 16.5, 16.6a-b, 16.7, 16.8 & 16.9a-b: Interior Designer: Bannenberg Rowell Design, Photographer: James Balston

Figure 17.1, 17.2, 17.3, 17.4, 17.5, 17.7, 17.9 & 17.10: Architect: BB Partnership, Interior Designer: Louise Bradley Interiors, Photographer: James Balston

Figure 17.6 & 17.8: Interior Designer: Louise Bradley Interiors, Photographer: Ray Main

Figure 18.1, 18.2, 18.3, 18.4, 18.5, 18.6, 18.7, 18.8, 18.9, 18.10, 18.11, 18.12 & 18.13: Architect: Flower Michelin Architects, Interior Designer: Todhunter Earle Interiors, Photographer: Ray Main

Figure 19.1, 19.2, 19.3, 19.4, 19.5, 19.6, 19.7 & 19.10: Interior Designer: Sybille de Margerie, Photographer: Fabrice Rambert

Figure 19.8 & 19.9: Interior Designer: Sybille de Margerie

Figure 20.1 & 20.12: Architect: ANARCHITECT

Figure 20.2, 20.3, 20.5, 20.6, 20.7a-b, 20.10b & 20.11: Architect: ANARCHITECT, Photographer: Ieva Saudargaite

Figure 20.4: Architect: ANARCHITECT, Photographer: Keir Harris

Figure 20.8, 20.9 & 20.10a: Architect: ANARCHITECT, Photographer: Chris Goldstraw

Figure 21.1, 21.2, 21.3, 21.4, 21.5, 21.6, 21.7, 21.8, 21.9 & 21.10: Interior Designer and Architect: Studio Reed, Photographer: Andrew Beasley

All other images are provided by the author.